Praise for *Facing Sunset*

Moving, powerful, and deeply personal, *Facing Sunset* is Patti Brehler's portrait of a life well-lived. A tapestry of her interwoven physical, spiritual, and emotional journeys—pursued both literally and metaphorically on her bicycle—Patti delivers a work of deep introspection and joyful inspiration. She shows us that we needn't fear what lies over the next hill, despite the wind or the rain, for we can only grow through our relationships with others, and through the time we spend in courageous solitary striving. Life is beautiful: *Facing Sunset* is beautiful. Bravo!

— Dr. Heather Neff, Professor of English and Director of the McNair Scholars Program, Eastern Michigan University. Author of eight novels, *Blackgammon, Wisdom, Accident of Birth, Haarlem, Leila: The Weighted Silence of Memory, Leila II: The Moods of the Sea, Blissfield,* and *Saffron Bloom.*

This book is written from the heart by a woman who's made peace with headwinds. Patti Brehler takes us for a ride, pedaling from place to place, person to person, decision to decision, and sometimes from trauma to triumph as she connects the dots of a life boldly lived. When I met her on the TransAmerica Bicycle Trail in 1976, she had tireless legs, a strong will masked by a disarming smile, and a contagious yen for adventure. Four decades later, *Facing Sunset* reveals that in addition to all of that, her most inspiring attribute is perspective.

— Rich Landers, former Bikecentennial '76/ACA bicycle tour leader. Author and co-author of four guidebooks, *Urban Trails: Spokane and Coeur d'Alene, 100 Hikes in the Inland Northwest, Day Hiking Eastern Washington,* and *Paddling Washington.*

Completed this book energized and inspired to live life to its fullest. As Patti takes us on her journeys, we are reminded again and again that "life is for living dreams." Her accomplishments are remarkable and the way they are interweaved with life lessons is inspiring and give one much to think about, even if not cycling across the country. The continual connections with others are uplifting, at this time especially. Her book reminds us of the goodness of people and the love that abounds. Her openness to be ready for an adventure will leave you inspired and wondering what you can do. Her feat to ride her bike until, like Forrest Gump—until I'm done, reminded me to listen and follow my heart.

—Becky Andrews, author of *Look up, move forward.*

Patti Brehler puts her heart and soul into this inspiring account of one woman's quest to recapture her youth and the spirit of freedom that comes from long-range cycle touring. It's a courageous story in every way, from the intimate details of her early life to the challenges of cycling the Great Plains, the Rocky Mountains, and the forests of the Upper Midwest. A must-read for anyone planning a long-distance bicycle tour, as well as a great adventure story for armchair travelers.

— Robert Downes, author of *Biking Northern Michigan* and *Bicycle Hobo*

Facing Sunset

3800 solo miles; a woman's journey back and forward

patti brehler

Lilith House Press
Estes Park, Colorado

ISBN 978-1-7353387-6-7 (paperback)
ISBN 978-1-7353387-7-4 (e-book)
Library of Congress Control Number: 2021904440

The stories in this book reflect the author's recollection of events and dialogue has been re-created from memory. Some names have been changed to protect the privacy of those depicted. All photos not taken by the author are used with permission by the Adventure Cycling Association, Robin Heil, and Karen Voss.

Cover and e-book design: Jane Dixon-Smith at www.jdsmith-design.com
Paperback interior design: Theresa Edkom
Editing: Elisabeth Kauffman at www.writingrefinery.com, Angela Mac, MaxieJane Frazier at www.birchbarkediting.com.
Cover and author photos: patti brehler

For Andy.

For Mom and Dad.

For that marvelous machine, the bicycle.

BICYCLIARY

R. Brooks

The bicycle is naturally singular,
Avoids ideology, does not confer
Virtue or signal the millennium.
It forces respect for equilibrium.
As technology, it falls somewhere between
The eggbeater and the sewing machine.
It is my size. It does not overbear.
It teaches attention to winds and the slope of the earth,
Loose stones and puddles, dogwood, a crooked path.
It asks me to get it from here to there
With competence and economy.
It is some pleasure to reply.

What you can do, or dream you can, begin it,
Boldness has genius, power, and magic in it
-John Anster, inspired by the words of Goethe

"What's the worst that could happen? I might die? Well, we are all going to die, no sense worrying about it.

What's the next worst thing? I'll be homeless and have to live under the bridge? No big deal, I have a sleeping bag.

What's another? That I might get attacked or raped? Well, then I might die, because I'll go down fighting."

-patti brehler

Contents

Forward

The author is an extraordinary, ordinary woman. Born and raised in Detroit, her education and experiences appear blue-collar-ordinary. She would have herself viewed that way, but this book will tell a different story.

Raised in the Catholic faith, the church's curious restrictions on women soon pushed her away. A talented student who earned the Valedictorian title at a large suburban high school, she spurned college to work in Detroit's manufacturing sector. Typical of her, she rose from performing rote assembly of minor car parts to journeyman machinist in an aerospace manufacturing company.

She has worked as a writer and photographer for a small, rural paper and rose to edit another such publication. At one point in her work life she "ran away" to work on the railroad. She successfully completed the strength-based training, but the male-dominated work environment frustrated her, and so her railroading career was short lived. Otherwise, she has asserted herself in the working world of men long before cultural change was there for support, and she earned respect from managers and (perhaps more significantly) her male co-workers.

During her early work life, she compartmentalized her employment to support her true passion: riding the open road on a bicycle. Before the ink was dry on her high school diploma, she was touring the roads of Michigan. Patti first rode her bike coast to coast when her young age still required written parental permission. Later, she competed in many cycling endurance events and set records in the USA and France, some of which stood for years after she no longer competed.

Unlike some endurance athletes, for her, once it was over, it was over. To fill the void created by the end of her competitive riding career, she turned to other interests. A lifelong cross-country skier, Patti plied the trails of where? The urban ski trails of Metro Detroit? Not a chance. She traveled beyond the reach of the electric grid into Canada to find a place of white magic. She also trained as an adult leader in Michigan's 4H Challenge Program, specializing in outdoor activities and winter survivor skills. And all through her life, writing about and photographing the miracles of life have been a constant.

Her friends (and I) affectionately describe Patti as more than a little crazy. She raced her mountain bike on the Alaskan Iditarod trail, in February. For Patti, camping in the winter snow with no more than a sleeping bag for warmth is fun. In the building which is now the headquarters of General Motors in Detroit, she set a world record for climbing stairs for twenty-four hours (just shy of 70,000 steps). A fund-raiser for the American Heart Association, her climb was an inadvertent protest of inattention to women's heart health studies.

While she never wanted to be the center of anything, when she saw a need, that's where she went. A trained massage therapist, Patti volunteered her services with Hospice. For a decade she raised puppies for Leader Dogs for the Blind. For several years she volunteered inside the Michigan prison system, helping selected inmates develop skills to raise Future Leader Dog puppies.

Throughout her life of competing and giving she was a friend and role model for women, young and old. She intended not to preach a message, but to live it. "You can be who you want. You can do what you want. You can lead the life you want." Finally, she saw it as her task to help both her parents on their journey from this life to the next.

One wouldn't think there was much room left in her life, but when Patti was in her late thirties, we stumbled into each other's lives. I had been a failure at marriage, twice; she had never felt a need for a lifelong partner. Credit me for keeping trying; credit her for taking a risk. That was twenty-seven years ago. For more than two-and-a-half decades I have watched and wondered at this extraordinary, ordinary woman. Her bike ride to Missoula, Montana and home again is a metaphor for her life. "I wonder what's down that road," she would say. Thankfully for me, I was down one of those roads.

Andy Andersen
2021 the patch, Lupton, Michigan

Facing Sunset

Leaving St. Ignace, Michigan, June 9, 2016.

Editorial note: to orient the reader to the author's structure, point of view, and tense in this book: the italicized chapter openings depict the author's growing-up memories written in third-person, past-tense; the narration of the author's ride is in first-person, present-tense; and other author's memories within said narration is delineated by a bicycle wheel, written in first-person, past tense. The italicized *Postcards From the Road* at the end of some chapters are actual Facebook posts the author wrote during her 2016 ride.

1

A Chance at Being Done

Thursday June 9, 2016
St. Ignace to Hog Island State Forest Campground, Naubinway, Michigan

His smell—cutting oil and grinding dust—brought the toddler running. "Daddy!"

The father eased himself to the shag carpet. Shirtless, his arms angled at the elbows like chicken wings. He clasped his grease-stained fingers and bowed his forehead against them as if he were praying. "Walk on my back, Pat."

The girl loved this game. Could she step from squishy butt to neck without falling? She waved her arms wide, one bare foot over each pocket of his navy blue shop pants.

"Ugh. Keep moving."

Toes curled at his belt. Tiny tot steps stuttered to catch balance, leaving white marks that turned red with passing. His tanned skin felt as hot as a cement sidewalk in June. The girl didn't realize she held her breath as she rocked her weight up either side of his spine.

"Yes…that's it."

Where his back widened, she connected. Moans and giggles mingled.

Warrior-child, dancing on daddy's back.

My heart skips. The first view of the Mighty Mac comes at a curve on northbound I-75, between mile markers 334 and 335—stark, carcass-white spires thrust above the forest canopy. The Mackinac Bridge, a five-mile span connecting the lower peninsula of Michigan with the upper (the UP), the longest suspension bridge in the western hemisphere, signals a leap from home.

Before cables connected the roadway to the 552-foot towers, we had to take a ferry across the straits. Mom cradled me in her arms on our way to visit her folks. In memory, her voice singsongs names of UP towns:

Naubinway, Manistique, Munising, Ishpeming, Negaunee, Ontonagon, Calumet. I loved her stories of growing up in Harvey, a small town east of Marquette on the shores of Lake Superior.

After the bridge opened to traffic the sprawled-on-my-back view from the rear seat window of our family's Chevy left me awestruck. Cables shadowed like a broken reel of Super 8 home movies. Sunshine washed either end of the span, and rain poured between the cloud-shrouded towers.

The Mighty Mac, a siren calling to adventure.

My adventuring started at age five on a bicycle; my first foray to follow the road was the summer of 1974. Fresh out of high school, my friend, Robin, and I took three weeks to pedal from Detroit to the base of the siren bridge. Restricted from riding across, we squeezed into the cab of a Bridge Patrol pickup, our loaded, ten-speed Raleighs resting in its bed.

I yearned to keep on, but we had a week to get Robin home for her brother's wedding. "Au revoir," I whispered to the Mighty Mac from a ferryboat hauling us back to the lower peninsula after a day's respite on car-less Mackinac Island. I'd ride back to the city with Robin, but I was not done living life from the seat of a bicycle. After resting a few days, I planned to empty my meager bank account and continue pedaling to Colorado, the place to be.

Sprawled out on my parent's hulking, Naugahyde reupholstered chair, I thumbed through the *Detroit Free Press*. An article caught my eye. "Look at this, Mom! These two couples are biking from Alaska to Argentina. When they get back, they're going to organize a ride across the country to celebrate the bicentennial year."

"Better than going on your own."

Scrap Colorado, I had a new plan. I wrote to Dan and Lys Burden and Greg and June Siple, the riding couples noted in the article, for information about their brainchild: Bikecentennial '76 (B'76). In the meantime, I went to work in a factory making electric trunk releases and motorcycle carburetors—money for new camping gear. The summer of 1975 I dragged my almost-fifteen-year-old brother, Jim, on a two-week shakedown tour to Lake Michigan and back. I was ready.

In 1976 I cycled 4250 miles from Oregon to Virginia. Afterwards? Forced home by a prepaid airline ticket, I still didn't feel done.

I got a better job in an aerospace manufacturing factory, with vacation time for shorter tours. Weekend rides found me exploring Ontario. One summer I rode six days across Wisconsin. I even took several more pedals to the Mighty Mac. They were not enough. I turned to ultra-marathon bicycle races, condensing weeks of miles into twenty-four hours or multiple days. Not the same. My heart lusted for an open-ended life on the road.

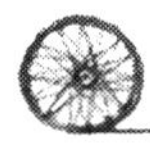

Today the sun shines end to end as Andy drives me and my loaded Tour Easy recumbent across the Mighty Mac. Today, on the north side, I begin my lifelong dream of "riding until I am done." I named this desire my "Forrest Gump" ride, after watching Tom Hanks run back and forth across the country in the 1994 movie by the same name. Hanks's character got up from his porch, and "for no particular reason" ran to the end of his road.

What stuck with me? He kept running until he was done.

Oh, to pedal where whim takes me, pushing rubber to the next turn. Bicycling: an elegant blend of body and machine is a force in my life I cannot deny.

"Another fine mess I've got myself into," I say to Andy on the north side of the bridge, clipping my right shoe into my pedal.

"Well, what can you do now?"

Through the lump in my throat I croak, "Give it my best shot."

He squeezes my shoulder. "Failing that you can call the ERD."

"ERD?"

"Emergency Recovery Department."

I give him a sidelong glance and smile. Poor guy. Hasn't twenty-two years of marriage taught him anything? The man has listened to me for years: "One day I will ride until I am done." He knows that, beyond a catastrophic physical failure, I will not ask for rescue. Can't fault him for trying.

The sun is high, west winds low in my face. It is June 9, 2016, and I have five weeks to ride from St. Ignace, Michigan to Missoula, Montana. And however long I want to get home.

I take a quick glance at the mirror sticking out from the left side of my helmet. Andy drops his hand after waving me off and turns back to the van. His head and shoulders droop. I imagine him saying to himself, "It's going to be a long summer." I am glad for the eight-foot shoulder on US 2 when tears break out, taking me by surprise, just as Andy did when he walked into my heart.

My back had been killing me for weeks. That's what a job that doesn't fit can do to a person. In mid-step, gimping through the office hallway toward the factory floor, I thought, *I gotta quit*. Before my foot hit the floor, all pain disappeared. I handed in my resignation letter.

My boss (president of the company) asked, "What will you do?"

"I don't know, it just can't be this."

"You've done a good job for us. I make it a practice to treat my staff to a thank-you dinner when they leave."

He came from work to pick me up. When I stepped in to control the exuberance of my two mutt dogs, he brushed me away. No mind to the black and yellow fur sure to adhere to his expensive suit, he dropped onto hands and knees to wrestle them.

Who is this guy?

This guy was Andy.

I worked at that aerospace manufacturing firm for seventeen years, fifteen of them as an O.D. (outer diameter) grinder. Looking for a change, I applied for and landed the company's new training coordinator position (with a requirement to go back to school for a degree). Even after I secured an almost half-million-dollar government grant to retrain the workforce, business tanked. Necessity shifted me from training coordinator to afternoon shift floor supervisor, not a job I wanted and the one that blew out my back.

Andy was incredulous. "Most people line something up before they quit. Or have a plan of some sort."

"I'll continue with school. I have a massage therapy side gig and hope to grow my client base."

He had no idea about the crazy person who worked for him. I admit to a bit of fun, at his expense, filling him in. Hadn't he ever met a woman like me? As president of the company, Andy represented the establishment; I was a factory-rat at heart. He said it himself: "I don't understand you." But my free-spirited approach to life and stories of my adventures piqued his interest. By the end of dinner, after he professed a "bucket list" wish to hike the Appalachian Trail (but no idea on how), we agreed I'd help him train for a week-long hike.

I called him the next morning. "You gave me an idea to start an adventure coaching company. Will you be my first client?"

A year later, my life forever changed. Andy and I finished a six-day hike from the southern terminus of the Appalachian Trail, came home to buy a bicycle store together, and got married.

And now I'm leaving him, the guy who made it his purpose to "make it so" when I first bridged taking another long bike ride. He pretended not to notice my warning, "The Pacific Ocean isn't that much further, I might want to keep riding west once I get to Missoula."

"I'll be fine. Just come back to me."

Unlike my mother, who frowned a tease: "Do I have to sign for you to go like you tricked me into doing before? If I knew back then you couldn't go without my signature, I wouldn't have signed."

At age twenty, I needed a parent's signature to sign up for B'76. I didn't trick her. Mom knew full well her signature would be authentic or forged. She signed. What she probably doesn't want to know now is she sparked the idea for what I am about to do.

Mom managed the family finances, filing every bill. In her organized way, she shredded every piece of paper with Dad's name on it after he passed. Seemed normal to me, she worked that shredder hard every year. It's what she said that struck me: "It's like he didn't even exist. I tell you, if you want to do something in life, you'd better just do it."

Did she have regrets? Or was she warning me? Was she aware of my itch for the open road and hinting for me to go?

Not a newspaper article this time, but a Facebook post about the fortieth anniversary of B'76 pricked my fire. Dad was gone, Mom appeared healthy, and I wasn't getting any younger myself. It was a perfect storm of opportunity.

I took Mom to her doctor's appointment, the doctor that signed Dad into hospice care. "Doc, I'm planning a three-month-long trip this summer. Think that's okay?"

He winked. "Sure."

Now I'm just doing it. Pedaling free. The open road a strip of circumstance luring me along the northern edge of Lake Michigan. Who cares about oceans when we have the Great Lakes? My tears are not enough to salt the sparkling expanse.

A highway sign: eighty miles to Manistique, tomorrow's destination. Tonight, I shoot for Hog Island State Forest Campground, a few miles east of Naubinway, thirty-five miles west of the Mackinac Bridge. A drumbeat lifts in the back of my mind to match my pace. Laurie Anderson's "Ramon," from her album *Strange Angels*, burrows a worm. A song I first heard more than twenty years ago, on a playlist my brother made for his dying wife. We played the tape while I massaged her.

With memories of helping Liz on her last journey, tears well again, a little harder as my thoughts turn to Dad. Funny how the memory of one loss causes memories of other losses to surface. I cry for the cycle of life, for those I've lost, for those I've yet to lose. I cry and pedal on.

At Hog Island, fifty rustic sites along the flat dirt road of the campground are empty. I choose a spot on the shore of Lake Michigan in hopes the lapping water might drown out traffic noise from US 2. It might be the exact spot where Andy and I stayed eight years ago. We were heading home on our bikes after visiting his son, daughter-in-law, and grandsons in Green Bay, Wisconsin. Another bucket list item for him. On our way we crossed the big lake on the Badger, a ferry from Ludington, Michigan to Manitowoc, Wisconsin. Our return took us north to visit Andy's brother in Marinette, Wisconsin before crossing the UP east, and a Bridge Patrol truck ride across the Mighty Mac.

"Hey," I said to Andy as we left his brother's. "What do you think about swinging north to Marquette first?"

"No."

I didn't think he could smell home from there, but he did. For me, the urge to keep going was stronger than ever. I still was not done.

At the end of this emotional day I should cry over my dinner too. Not many choices in the scruffy convenience store in Epoufette a few miles back. Turns out a can of tomato paste mixed with water and spices does not make a tolerable sauce for the pasta stashed in my kitchen pannier. The disaster reminds me of a phrase Andy stole from a Garrison Keillor character whenever I concocted a meal that wasn't quite great. "I ate it, didn't I?"

Tucked in the same down sleeping bag I used in 1976, I toss and turn. Am I up to this? Can my body handle the stress? Do I really think I'll keep riding after Missoula? Andy has always encouraged me to go; why did I need the reunion as an excuse? My mind races against my heart like the moments leading up to the start of a race, or a new job, or a first date.

By 2:00 a.m. my bladder needs attending. Now I need the gumption to give up the warmth of my bag and hike to the vault toilet. Ah, well. Zip. Or rather, unzip. Tent door, vestibule door. Zip again to close against mosquitoes. I stumble out.

It is not the chilly UP air that forces a gasp. The Milky Way shouts its existence in reflection over calm Lake Michigan waters. I hover among stars at the brink of earth and water, existing in human form this brief moment of time, occupying a nano-space of no consequence, at once all and nothing.

STATS (From my CatEye cycle computer.)
35 miles, max speed 37.5, ride time 2:48, average speed 12.7, TOTAL 35

Dad and me.

2

Yes, I am Alone

Friday, June 10, 2016
Hog Island to Indian Lake State Park in Manistique, Michigan

The girl was yet another sister (the third) to Rick, the oldest. He dragged her around as if she was a brother. "Good thing you're smart," he said. By age two, she knew right from left.

Rick and the neighborhood boys built plywood go-carts to race. They nailed metal roller skates to two-by-fours for axles and attached the front steering axle with a large bolt and washers through an oversized hole. Drivers steered with their feet and pushers ran behind with a broom handle wedged against the seat back.

Except for her short legs, the tiny girl was perfect. Rick nailed two lengths of clothesline to the front axle for her to steer. All he had to do was yell "left" or "right."

Wheels clack, clack, clacking on sidewalk cracks, the brother/sister team was unbeatable.

Gentle mourning dove coos wake me back to earth. A tap on my FitBit glows the time at 5:41 a.m. Good, I want an early start. Pancakes and peaches fill the hole from last night's dinner, but even with the early up it is almost 8:00 before I get everything packed. Ready to roll, I skate free. No forest ranger ambles up looking for the $13 camping fee I should have dropped through a self-serve slot in a steel post at the entrance. With nothing smaller than a $20 bill, I took a chance. Happy to pay, disinclined to deplete my limited cash with a tip.

Andy was right. "Use the debit card, but you'd better bring some cash, too. It might come in handy."

Forty years ago, I bought $1000 in traveler's checks as a safe way to carry money for extra snacks, and souvenirs and postcards to mail home. Using cash these days is like shooting film. Who does that anymore?

The St. Ignace Truck Stop Restaurant for one, it turns out. No credit cards accepted. I paid for our lunch and the bridge tolls yesterday, as Andy brought no cash of his own.

After an hour of riding I pull into a rest stop teeming with vacationers. Lake Michigan waves roar wild. The only water I'm interested in comes from the water fountain—to fill my bottles. Good fortune, underneath it is an electrical plug. I worry about keeping my iPhone charged. After reading tips on bicycle touring blogs about "guerrilla" phone charging (vending machines often have open plugs behind them, for example), I decided against bringing a portable charging pack.

I linger long enough for a half-charge. When I return to my bike, it is lying on its side. Was I careless leaning it against the walkway railing? Did someone knock it over? Nothing seems amiss. A gray-haired couple grip the handrail as they shuffle to their car. They pause. "Where are you going?"

"Missoula, Montana."

"What for?"

"A fortieth anniversary party. In 1976 a bunch of us rode across the country to celebrate the bicentennial year."

"That's a long way just for a party."

I grin. "It might be quite a party. More than 4100 people cycled that summer, from either coast or shorter portions. I thought it'd be neat to ride there, while I still can!" I don't bother to tell them how the Burdens and Siples kept on with their B'76 organization and developed bicycle routes all over the country, or that they eventually changed its name to the Adventure Cycling Association (ACA).

"Are you alone?"

"Yep."

The man's eyes widen. His wife breaks into a Mighty-Mac grin. "Good for you!" I wonder if she has ever yearned for a similar adventure. Then I wonder if I'm setting myself up. Should I not admit I'm alone? Maybe I should answer, "For now. I'm meeting up with some friends." Not a lie, I am meeting fellow B'76 riders in Missoula.

I point my rig west onto US 2.

As if on cue, "Ramon" starts right up. Curious. If the song hadn't popped in my head yesterday, I would say last night's stars bring it to mind today—Anderson's "kerjillions of stars" rising like angels. I relax into a light tailwind and a gentle, familiar road. My muse on her lovely lyrics, about connection and not knowing, is interrupted by an Eastern Massasauga rattler stretched halfway across the shoulder. Its head tickles the gravel edge, there is no flat spot on the body or evidence of body fluids. *Is it dead?*

I swing a wide berth.

Halfway to my destination at Indian Lake State Park west of Manistique, I stop at a lone gas station for lunch. In line to pay for a red Gatorade and Lays potato chips to compliment a peanut butter sandwich, two women ask about my bike and where I'm going.

"It's a recumbent." I launch into my anniversary spiel.

"Are you alone?"

I try out my new response. "For now. I'm meeting some friends."

"Where are you meeting them?"

I can't lie. "In Missoula."

"Well, you are alone then."

Scrap this idea.

The road turns south and with it the wind. A crashing in the woods is louder than the wind noise now blasting through my helmet. A deer bounds, escapes whizzing traffic, and disappears. I glance right, expecting a second. What I see instead makes me wonder if lunch hasn't kicked in yet. Crouched in the overgrown yard of a tumble-down cabin is a green frog as big as a Volkswagen. The frog faces the house, poised like a football center about to make a snap. It wears painted-on blue jeans and what appears to be a real black fleece vest and knit winter hat.

Not much further on I pass a discarded oil furnace tank painted white and black to resemble a cow, with a stovepipe neck and bucket head. A bearded, long-haired dummy-man, wearing a weathered leather coat, leans against it. He has aged since Andy and I saw him eight years ago. I swear his head turns slightly to watch me pedal by.

Only my second day on the road. Should have taken photos as proof I'm not hallucinating.

At a gas station in Gulliver for a potty break, two men drive up and park. The driver helps a frail woman get out of the back seat. She's dressed in a light cardigan and long pants of purple, yellow, and green pastels;

a bejeweled butterfly broach flutters over her heart. A bit disheveled, she looks like a wilted, late-spring lilac bush. The men are clean cut, wearing khaki shorts and polo shirts. Are they brothers? Or lovers?

The woman totters past me. “I like your bike.” The man straightens her cardigan, she mumbles something else.

“What did you say?”

She flinches. “She took my clothes!” The man hurries her into the station.

The woman’s swift reversal reminds me of Mom, how her sharp criticism has inured me and yet there are times she catches me off-guard with encouragement. Years ago, when financial difficulties at the bike store forced Andy and I to sell our house, I showed Mom and Dad the low-income townhouse Andy found for us. I expected some snide remark. Instead, she said, “That looks nice!”

I think of Mom living alone in Parkplace Heritage Village and the elderly residents I’ve gotten to know since she and Dad moved there two summers ago. The senior living apartments are a wonderful place with meals and cleaning, and activities if so inclined. When Dad needed more around-the-clock care, they hired aides from a home-health-care provider on site. Gives me peace of mind knowing she’s in a safe place, even if that place isn’t with me or my other siblings.

By 2:00 p.m., I cruise up to Jack’s Fresh Market in Manistique feeling strong. Should I push on to Escanaba? That would more than double my mileage today. Better not, I need to ease into this.

The market is a far cry from yesterday’s dinner shopping. Overwhelmed, I crave everything. I settle on a skinless chicken breast and a softball-sized head of cauliflower, bound to be 2000 times better than last night’s fiasco.

STATS

61.14 miles to Indian Lake State Park, max speed 32, ride time 4:36 (lots of stops), average speed 13.3, TOTAL 96.91

Indian Lake dinner

3

Headwind Queen

Saturday, June 11, 2016
Manistique to Escanaba, Michigan

The father built the rink in the backyard because his wife loved to ice skate as a kid. He stamped snow to form dams. Evening after evening he sprinkled water to freeze in layers overnight until a smooth, thick surface hardened like glass.

The mother twirled on the ice, her face framed by the fur on her hood. The girl thought the image magical; she longed to glide so effortlessly. Bundled head-to-toe she faltered, barely able to balance on her double-bladed skates.

"Take my hands, Pat." Her father's strength and confidence surged through her mittens. It felt like flying.

"Eleven more days." Andy's voice in my ear, me still snug in my sleeping bag. Indeed, a cell phone is nice to have along. In 1976 I hoped my folks would accept charges when I called home. Could I even find a pay phone today?

"You're counting down the days?"

"Yep." Days to my self-imposed deadline. He knows my Forrest Gump dream ride needs to be at least two weeks long—one week to beat myself into shape and the second week to adapt to the touring lifestyle. He must realize I'm committed to reaching Missoula, having paid $76 for the Friday night reception and Saturday night dinner. After that, well, could I be done? Or, like Forrest Gump, who "just felt like running," would I keep on riding?

By the time I leave Manistique it is 9:30 a.m. I find a paved side road leading back to US 2, the main east/west thoroughfare across the peninsula. A short downhill curve swings past a two-story farmhouse.

It has no siding except for a thick growth of ivy. A dusty white patio loveseat on the wraparound porch catches my eye.

Is someone lying there? I brake, giving up momentum to verify what I think I see. Yes. A skeleton lies shrouded in a dank sheet, as if the lady of the house died there waiting for her lover to return. Paved roads might be rare in the UP, but curious yard art seems to be a thing.

At the turn onto the highway, a gust of wind catches my fairing and again I am glad for the wide shoulder. Looks like this headwind is here to stay. Good thing I made peace with it years ago.

In 1985, my tandem partner, Lou, and I entered the John Marino Open in Illinois, a Race Across America (RAAM) qualifier. We needed to complete 700 miles in seventy-nine hours. We managed 500. Even with the power of two presenting a single front against a thirty-five miles per hour headwind, it was a battle to manage thirteen. I'd be thrilled with that speed today. On a tandem, it was pathetic.

Ignoring my lack of double-digit speed, I pedal on and smile at the memory of the two of us sprawled in a grassy ditch, the only place of refuge from the incessant wind. Our bike didn't crash. We did.

I pedal on and try to muster the forty-years-ago enthusiasm fighting a horrific cross wind in Wyoming during B'76. Eight of us hung together for solace, leaning sideways to keep upright. Our unstable pace line wavered. I took the rambling lead. "Welcome to the Wind River Reservation! Keep two inches between the wheel in front of you and you'll be fine. I'll be your tour guide today, so relax and enjoy the ride." I sang into the gusts long enough to displace my riding partners' wind worries and for them to shout, "please stop singing!"

In Gladstone, fewer than ten miles from my intended destination of Escanaba, the smell of broasted chicken lures me into a gas station convenience store. I don't resist.

Leg muscles squeal as I ease onto a wood-slatted bench outside the store. Leeward, I feel the full front of heat. Wind against a sweaty face makes eighty-five degrees feel not-quite-so-hot. Thirty-two ounces of red Gatorade disappears. I devour two greasy chicken wings, dig into my feedbag, and polish off yet another peanut butter sandwich, a banana, and a few pieces of beef jerky.

I force myself back onto a wider, but busier, US 2. Has the wind lessened? Maybe it shifted, maybe I just needed a break, maybe I can ride beyond Escanaba. I cruise past the Pioneer Campground where Andy and I stayed on our way home from Green Bay.

The shoulder disappears and a steep hill reminds my quads of miles already ridden. A whirlwind of sand along the curb dismisses thoughts of passing through this busy town of over 12,000. Here, the UP State Fairgrounds has camping. A few RVs dot an arid open field, a too-long walk to the restroom and shower buildings.

"I think I'll head back to Pioneer."

Andy and I must have stopped here midweek, we had the shady place almost to ourselves. Today the modern campsites are full. I'm sent to the tents-only, pit-toilet, partying-youth section at the rear end of the property. Ah, well.

Hot showers are worth the excursion to the modern side. What a delight. Three sinks fill only one end of a long counter, so I perch at the far end, plug in my phone, and check emails and Facebook. I'm surprised how I'm enjoying this connection. Thought I'd stay more unplugged.

A woman comes in with a curly, red-haired toddler clinging to her hand. He startles. "Mom, there's a boy in here!"

"No, it's a lady. She just has short hair, like Grandma."

That satisfies him and he skips with her into a stall. I chuckle. *Guess my weight-loss getting ready for this trip has sent my boobies packing!* Been a long time since I've been mistaken for a boy. Not that I ever minded. Projecting a male image confers a sense of security.

Word of the night: trains. One rumbled by beyond a regimented stand of red pines as I set up camp. Nice to know the track was there before I went to bed. It reminded me of a stormy night spent in a tent with my younger brother, Jim, on our two-week tour in 1975.

Our tent was on a rise under an old oak tree next to a long, tall hedge in Ithaca's City Park. The night brought a rock 'em-sock 'em thunderstorm. We stayed dry, the storm passed, and despite sirens, we slept. Suddenly, the earth shook me awake. A tornado? Nope. A train. I didn't remember seeing tracks. Did the rain wash our tent from its perch? I was afraid to peek out, surely if I did, I'd see a cartoon train bearing straight for us.

I grabbed my brother. “It’s been great knowing you kid!”

In the morning light our tent was still where we planted it. A train did not mow us over, the tracks ran behind the hedgerow. Weary-eyed, we walked a short block to Main Street for breakfast at the town’s only café. We weren’t the only patrons craving coffee, it brimmed with local firefighters, exhausted from a night of fighting blazes.

STATS
57.19 miles, max speed 26.5, ride time 4:59, avg speed 11.4, TOTAL 154.1 WINDY!

4

Hitchhiker

Sunday, June 12, 2016
Escanaba to Crystal Falls, Michigan

The girl, half as tall as the fishing pole that wavered in her sweaty hands, stood trembling at the edge of Pontiac Lake. She did not want to swing the hook and bobber into the water. What if a fish snagged the worm and pulled her into the murky water? She wasn't strong enough.

"It's ok, Pat," her father said. "I'm right here. Let me help."

I reach to sweep a pinecone from my collapsed tent. The hitchhiker bounces into a fold in the fly.

"Wait," Pinecone says. "Help me get away from my family."

I rock back on my heels of imagination. "What's wrong with your family?"

"They laugh at my dreams."

"What dreams? And why would they laugh at you?"

"I long to see the world. They say I should plant roots or die."

I sense a presence and peek over my shoulder at looming pines. "Maybe they just want to protect you."

"Protect me? I'm suffocating."

Are those faces lurking behind that scaly bark? I cradle Pinecone in my hand. "I'm sorry. But you can't come with me. Your family is right. You need to grow roots to live."

"Then I don't want to live."

"Oh, Pinecone. Can't you see your potential? You family is only partly right. Be patient, and you can still realize your dreams. One day you will grow sky-high majestic. Imagine the world you will see then."

If anyone else is awake in the campground to witness my conversation they might think I'm losing it. Maybe I am, a little. Ah well. I pitch Pinecone as far as I can into the sunlight. Pine needles rustle.

Northwest out of Escanaba, I gain an hour crossing into the Central Time Zone. While the roads ahead are new to me, a forty-five-mile stretch of Michigan State Highway 69 reminds me of my home county of Ogemaw. Smooth pavement rolls ribbons between wrapping paper squares of potato fields, hardwood and pine forests rise like gift bows.

A breeze caresses my right cheek, swirls across my chin, and brushes the left side of my jaw with a flourish. After a mile-long climb, I peek at my mirror. Clear traffic. I hang on to enjoy the thirty miles per hour ride while I can. Until. Movement freezes as if in a scene from the *Matrix*. A monster bumblebee zooms in a trajectory that will intersect with mine; in slow motion, I dodge my head. A near miss and normal time returns.

Pedaling as meditation. When my brain isn't empty, I try things. What if, at the approach to the next ridge, I drop into the granny (the smallest gear on the crank) *before* clicking into the largest (and easiest) rear cog? Ah, better for smooth shifts into low hill-climbing gears. No stalling. After fifty-five bike-years, how did I forget this? The only tiff Robin and I had during our after-high-school tour was over how to shift.

There's a reason for multiple gears on a bicycle, and the way Robin always left hers in a high one drove me crazy. "You should pedal quicker, Robin. Don't be afraid to shift, the gears are there to make it easier."

She ignored me and cranked harder. At the approach of a hill, Robin threw her whole body into the effort.

"Push your left lever forward to get into the granny gear. Like this." I shifted to demonstrate, and my legs hopped into a spin. She stared forward.

I waited for her at the top. With her chain still in the big chainring, she struggled to turn the crank. "Please. Just try shifting the way I tell you. If it isn't easier, I'll shut up and you can shift any way you want."

"Fine."

I eased in behind her. "Leave the chain in the big ring now. Your pedaling cadence should be between sixty and eighty rpm."

We raced the gravity pull until the roller-coaster road rose again.

"Now, drop that big ring to the small before it gets too hard." Back then we didn't have triple chainrings with a super-low granny-gear.

Robin's legs jerked until the rise slowed her cadence. "Now pull the rear shifter toward you, that's it." Her rear derailleur leaped into a slightly larger cog. Her legs spun. Before they slowed, "Again, the same thing."

We swept the hill without grunts. She felt it. And that was that.

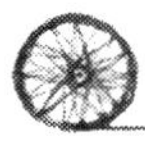

A gentle downhill run delivers me to my day's destination, a grand sweep right into Runkle Lake Park and Campground, sparing me a spectacular climb into Crystal Falls. Are there falls? The town itself cascades below the Iron County Courthouse, visible from a mile and a half away. I hope the grocery store is nearby.

The camp office is a tiny weathered cabin with a squeaky screen door that slams behind me. A desk sits empty, muffled noises waft from a second room. "Just a minute." A man's voice.

Caretaker Harry is a seventy-year-old veteran who winters in Arizona. Not much taller than my five-feet-two, his bald, round head holds forth a smiley face.

"I can't see charging you the full $12 for a tent. How about ten bucks?" He writes my registration in a spiral notebook and re-enters it into a computer. "Give me a minute and I'll print you a receipt."

"I don't need one."

"We've got all this new stuff and I want to try it out. By the way, I have an Adventure Cycling group coming in here in August."

Since 1976, the ACA continued mapping almost 50,000 miles of bicycle-friendly routes around the country. Cyclists buy maps to plan their own tours, as I did, or sign up to take theirs. With the ACA, one can ride across the country, part of it, or even for a weekend, and fully supported, self-contained, or inn-to-inn.

"Where's the nearest grocery?"

"On the west side of town."

Up and over the hill. Ah, well.

Other than one deserted tent (and RVs in a separate area behind Harry's office), I have the park overlooking wake-free Runkle Lake to myself. What a deal for ten bucks, electricity and hot showers. And look, two quarters on the picnic table. Good luck for me tackling a grocery run?

Before I head out, Harry putters over on a golf cart and gestures to the Sierra Designs tent Andy bought for our hike. “I’ve never seen a tent like that. Most of the cyclists’ tents are so small I can’t believe you could even turn over in it.”

“Yes, it’s an old one, and I like the extra room. Still works fine.”

A young teen rips through camp on a minibike, spitting sand along the beach. Harry intercepts him. The boy leaves at a slower speed.

“He’s not supposed to ride here,” Harry reports. “Don’t you worry none, I keep this park safe.”

I don’t worry none, even though my grocery run turns into a walk. Forced to dismount, I push my bike up each side of the hill in town.

An easy rain soothes a comforting sleep. I feel a dribble, dribble on my cheek. I roll. Then splat. Square between the eyes. *I swear, I didn’t kidnap your pinecone,* I half-mumble in that drowning place between sleep and wakefulness.

Splat!

Wait, I’m not dreaming. Where’s my headlamp? A puddle grows in one corner and seeps toward my food bag. I pull it away. From the outside, I imagine my tent presents a light show, headlamp beam sweeping inspection. There, water droplets gather along the rain fly seams, seep through the screen roof, and drip. The fly is waterboarding me, not some government agent.

I stash what I can into my waterproof Ortlieb panniers and pile everything to where the tent floor is dry. Should have sealed the seams. If I can’t find seam sealer, I’ll have to buy a tarp large enough to cover the tent.

A gypsy, that’s what I’ll look like.

STATS

84.17 miles (2.5 miles into town and back for groceries), max speed 30, ride time 7:09, TOTAL 238.27

Robin and I depart Detroit, MI, 1974.

5

Hills, Hills, Hills

Monday, June 13, 2016
Crystal Falls, Michigan to Star Lake, Wisconsin

Saturday bath night.

Her father knelt on the rug and scrubbed Pat's back, a pack of cigarettes rolled under a sleeve of his white t-shirt. She held her nose as he leaned her backward to wet her hair.

She squeezed her eyes closed. He squirted a dollop of shampoo onto her head and kneaded it in with his fingertips. Head lolling side to side, she wished he'd use his fingernails instead.

Afterwards, toweled and warm, he sat on the toilet lid and braided her long brown hair. Tighter, she thought, pull it tighter.

The morning is cool enough to don tights and fleece. A half-wall, stone pavilion offers pancake-cooking shelter from the still-spitting rain. Inside is a fireplace fit for a giant. From beam to beam, a house wren flits a visit, now on the half-wall, now on a nearby picnic table. It chats as if asking for a share.

"Beautiful morning despite the rain, no? I need to find seam sealer though, I don't like getting wet in my tent." I flick a bite of pancake its way. The bird darts away. "What? Don't like pancakes?"

The rain quits soon enough but not in time to avoid packing a wet tent. Ah, well, Wisconsin today!

Only eight miles past Crystal Falls, I sigh back on the bike after walking up the fifth hill of the morning. A car slows from the other direction. The woman driver lowers her window and nods her head down the road. "I just saw a bear. You might want to be careful."

Careful? On a bicycle? When our B'76 group pedaled into Yellowstone National Park, a sign warned: "Keep Windows Closed. Bears." Not something you want to see when you don't have windows.

Good thing we never saw any.

Andy's daughter, Jen, made-up a "Hey Bear!" song to sing on hiking trips out west. I can't remember her verses, so instead I sing aloud with Laurie Anderson, "Ramon" still looping in my head.

I do not walk another hill for several miles.

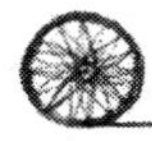

It would have been fitting for me to leave our home in northeastern Michigan on May 27, the 1976 date that our camping group dipped our rear wheels into the Pacific Ocean, but Andy turned seventy-two on May 30. The man is so accommodating of my crazy ideas I couldn't bear leaving before his special day. I targeted the first weekend in June.

Not to be.

One night Andy rolled out of bed with a groan and headed for the bathroom. It was seven weeks after his March hip replacement surgery and although he managed fine without me, I alerted to his movement. I sunk my head into my pillow and awaited his return to the warmth of our down comforter.

A loud bang hauled me fully awake. Did he fall?

A clatter outside halted my leap after him. Behind the window blind the moon neared new, but dark-busting light from across the road was enough. A black bear ambled through our English garden, only a narrow path away from the back porch, looming as large as our Hawken outdoor wood boiler.

"There's a bear in our yard!"

"There's a bear in my back!" Andy moaned. I knew things were serious when he said yes to my question, "Do you need me to take you to the ER?"

The bear took out our bird feeders; the bear in Andy's back was a kidney stone. A few weeks later the stone passed, but a call from the doctor revealed a new problem. The CT scan picked up two abdominal aneurysms. They scheduled an ultrasound for June 8 with a vascular surgeon appointment a week later.

I put my departure plans on hold.

Andy wouldn't have it. "You should just leave. It's unlikely they'll do anything, and if they do, you won't be far. You can come back."

"I can't leave when there's a chance you might need a procedure." The best news would be if the aneurysms were small (thank you Internet)—nothing to be done except watch them. However, if the aneurysms were large, it meant surgery. "They must know something from the CT scan." I hoped when he went in for the ultrasound that we could see his doctor.

In the meantime, I researched train schedules. I could pick up Amtrak in Flint and take the train to Chicago, and from there take the Empire Builder to West Glacier. Amtrak offers bicycle boarding, but they restrict recumbents. That meant packing and shipping my bike and equipment. From West Glacier it was only a few days ride to Missoula. If Andy needed surgery, I could still make the celebration and take my Forrest Gump ride home afterwards.

Andy called the surgeon's office to demand he talk to us when he came in for the test (oh to be a fly on the wall). "You don't understand. My wife is scheduled to leave on a summer-long bicycle trip to Montana, but she won't go until she knows I'm okay. I need the doctor to talk to her before my test."

He convinced them. The surgeon came to the waiting room. "I'm sure he's going to be fine. The most likely thing is that we'll just need to check them every year. The test today will give us a baseline. You can leave on your trip."

I was satisfied.

"I'll drive you over the bridge tomorrow," Andy said, giving me a few days jump on a week-behind-schedule.

Hills, hills, hills. And more hills.

Why do I do this? For a sense of adventure? For the satisfaction of focused effort on a single-purpose goal?

Peak experiences. I've had a few.

Crewing for another friend's tandem team in the Race Across America (RAAM). Everything I learned about riding a bicycle long distances put to use getting them from California to Florida in eleven days. Massaging my brother Rick's wife for six months as she died from breast cancer that metastasized into her brain. My time with her led me to volunteer with hospice, giving massages to caregivers and the dying. And Dad, helping him pass, a task I grew up expecting I would one day face.

Efforts like these extend my instinctual side. I am alive, in line with the universe. Politics, the economy, controversies—nothing else matters more than doing what needs to be done. Surviving the day. Everything here and now. And things basic, like on this ride across the country, finding safe passage, food, and a place to rest.

This ride, at first glance more physical than emotional, brews of something greater. Less risky than adventures like those of Lewis & Clark, or the Donner expedition stuck for a winter in the mountains, or Amundsen reaching the south pole, this ride, this ride is perhaps a quest?

Hills, hills, hills.

The day is dim when I stop at the Star Party Store near the state park, my night's destination. The little store offers boat rentals, propane, and firewood. Hunting and fishing gear, bait, tourist t-shirts and souvenirs, candy, chips, ice cream, frozen pizzas and other groceries, pop, beer, and liquor overflow shelves.

What they don't have is seam sealer. I'll keep things stowed in the center of my tent and hope for no rain. Ah, well, at least I never saw the bear.

I discover a new boxed meal, Hormel Compleats Chicken Alfredo. Meant for a microwave, I figure to heat the contents in my non-stick pan. At 320 calories and sixteen grams of protein, how can I go wrong for less than three dollars? I stock up on Kind bars and pick out five postcards for a dollar.

"Do you want stamps?" The cashier is barely visible behind a narrow stretch of counter crowded with merchandise. I forget my helmet is still on my noggin. Debbie, actually the proprietor, asks about my ride.

"We get a lot of Adventure Cycling groups coming in," she says after I tell my story. "And they don't realize how campgrounds here fill up during peak summer season. One group came in looking for a place. Their leader quit and a young British woman took over. She told me she had no clue what to do. A man from her group barged in and demanded 'real fucking food, not this freeze-dried crap!' It was an ugly situation. The lady apologized. There was no room at the campground. I called my dad, he came in and said, 'Who's the guy that needs to get away from this group for a while?' He ordered him into his truck, put his bike in the back, and took him home. My mom barbequed dinner for him and the group ended up spending two nights camped on their lawn."

Must not be peak season yet, the West Star Lake campground is not crowded. My wet tent dries out on a ridge overlooking the lake. Mom FaceTimes me and I turn my phone to show her the serene view, tall hardwoods and pine trees as far as I can see.

"Where are you? Those look like Wisconsin trees."

Funny, that's what I thought.

STATS
71.92 miles, max speed 34.5, ride time 6:43, avg speed 10.6, TOTAL 310.36

Four-and-a-half days to cross the UP. Goodbye for now, Michigan!

6

Chores

Tuesday, June 14, 2016
Star Lake to Lake of the Falls, Wisconsin

"Again. Again. Again!" At age five, Pat aimed a tiny two-wheeler bicycle down the gentle driveway slope of her family's post-war, west-side-of-Detroit bungalow. She pushed off. Tennis-shoed feet dangled as rudders, when she felt the fall she tapped the ground to regain balance.

To avoid careening into the street, she fought to turn left at the sidewalk and rocketed wide to crash land on the soft curb grass. Face scrunched in deep focus, she untangled herself, grabbed the handlebars, and stomped up the driveway.

"Again." She mounted. "Again." She rolled. "Again." She made the turn.

The act of pedaling with both feet came quicker than learning to control the fall. When she raised her second foot to plant it on the pedal, she escaped gravity and felt freedom for the first time. She kept on pedaling.

The world was never quite as small again.

Six days in and I am still learning the best rituals for camp chores. For example, flame residue blackens my WhisperLite backpacking stove. If I pack it and the heat diffuser plate away and then wash dishes, my hands get clean at the same time. Then I realize, no wonder I'm burning so much fuel. The diffuser is for baking, part of the oven kit I left behind after a pizza dough fail on an overnight shakedown a month ago.

I toss it aside. This morning's water boils in no time.

The thirteen-mile cruise into Boulder Junction (Muskie Capital of the World) is a rollercoaster, hills balanced so downhill momentum gives speed enough to roll up and over. At the entrance to a busy flea-market, a 105-pound Bernese Mountain dog greets me with slobbery kisses. I stroll between booths filled with crocheted towels, handmade jewelry, tie-dyed t-shirts, and wrap-around skirts.

Nothing catches my eye enough to warrant adding more weight to my rig. Funny. In 1976 I strapped on all kinds of souvenirs. Somewhere in Kansas, I found a rectangular sheet of gray metal in the ditch. The off-set printing press plate sported a black-and-white photo of a welder with the word "Progress" printed across the top. I scratched my leg more than once getting on and off my bike before I found a post office to send it home. Later, I welded a metal frame for it, and to this day it hangs in our basement laundry room.

Picturing that sign hanging in our basement reminds me of my dirty clothes. And there, the next street over—a Laundromat. First time washing clothes this trip. An out-of-order sign on the restroom prevents me from changing into my Pearl Izumi short shorts and a tank top—my substitute bathing suit and only clean clothes. A man waits in front of a clothes-filled dryer.

"They put that sign there to keep the tourists from using it," he says. "I'm seventy-seven and I've seen a lot in my life, so if you want to just change, I'll avert my eyes."

A younger man sitting nearby glances our way. The place is floor-to-ceiling windows. I politely decline. "That's ok, it's supposed to rain tomorrow so I'll just wear these again."

What luck, the touristy town is home to Northern Highland Sports, a fishing and hunting supply store. This is more fun to shop than the flea-market. Seam sealer? Score! And can't resist a water bottle with a built-in filter to remove viruses, bacteria, parasites, and even the iron taste of well water. Don't want a repeat from my ride in 1976.

We were sick in big sky Montana. Our group was one of several that came down with dysentery. Stomach cramps and diarrhea crippled us while pedaling and sleeping bag accidents caused middle-of-the-night rinses. By the time we got to Yellowstone, the illness forced an extra rest day.

Health departments in Idaho, Montana, and Wyoming stopped cyclists for stool samples, struggling to find the cause. A rider in our group did not get sick; that he stayed in a hotel one night in Idaho while everyone else camped in a park helped pinpoint the source. A broken dam flooded a sheep farm and contaminated the park's well.

When you got no cover and you gotta go, rain ponchos made impromtu shelters for squatting. Too miserable myself to appreciate the humor, I later wished I took photographs of the yellow "porta-potties" dotting the Bitterroot Valley.

Less than ten miles from tonight's Lake of the Falls County Park, I stop at a Subway in the town of Mercer. Thanks to a gift card from my friend, Tammy Bartz, no cooking for me tonight. Tammy and I volunteer with Leader Dogs for the Blind, raising puppies to become guide dogs. We also advise inmate raisers in the non-profit's prison puppy initiative. When we travel to the prisons in the program, we often stop at Subway. Threatening clouds can't deter my enjoyment with a six-inch tuna sub, SunChips, and a Dr. Pepper on a raised deck outside the restaurant.

Dr. Pepper. Favorite drink of my eldest sibling, Rick. The Christmas I was in seventh grade, he gifted me J.R.R. Tolkien's *Lord of the Rings*. My legs went numb reading the trilogy at the kitchen table over the holiday break, sucking salty pumpkin seeds (another gift), and washing them down with the cherry-flavored elixir. *Why did they even come out with a "Cherry Dr. Pepper" years later? The original tastes cherry to me.*

I raise my pop can to Rick and Tolkien. Tolkien's worlds are always close to mine. When Robin and I trolled the forests of northern Michigan in 1974, we hollered, "Has anyone seen the Ent-wives?" Eight readings later, the trilogy remains my all-time favorite book series. And while I do not often drink pop, when I do, it's Dr. Pepper.

Lake of the Falls, nestled deep in a hardwood forest, is self-serve. Campsites are spaced along the Turtle River between Lake of the Falls and the Turtle Flambeau Flowage. Two sections with electricity are more expensive than one without, so rustic it is. This time I pay.

One deserted tent sits forlorn near a roaring fall at the rear of the campground. The noise tricks me, it sounds like a storm approaching. I much prefer the robin that serenades from a nearby stump as I set my tent up under a sprawling oak tree.

The shower building by the modern sites are only a tenth of a mile walk away. Best laid plans are sometime for naught. The showers need quarters, however many I have are in my tent. And the lights don't come on, so I can't even charge my phone. Added insult: thunder announces rain, maybe it wasn't the waterfall. I hustle back before my shower is free.

No cell signal means no FaceTime with Mom or phone call with Andy. At least I talked to him earlier in the day. "Tomorrow is your seventh day," he said, still counting down my two-week decision target, hoping I'll decide I'm done.

I admit to doubts. I need a better night's sleep, the time change is an unexpected challenge, light earlier and not as light later. A mosquito buzzes. Is it in the tent? I swat. No, it is between the screen top and the fly. A second buzzer joins in, looking for meat. Ah, well, here's hoping the seam sealer holds.

I'm not stopping before Missoula.

POSTCARD FROM THE ROAD (Written on 6/14/16 but posted on Facebook on 6/15/16, as I had no cell service at my campsite.)

What makes a road "good?" Pavement smooth enough to trust letting loose on downhills? Hills you can roll without dropping into the granny gear? Just enough traffic to give you a boost in a head wind? Passing hoards of purple, blue, pink, and white lupines growing on the side of the road?

All of that. I needed a shorter (forty-seven miles) "recovery-ride" yesterday. Almost twenty miles of it was on a curvy, paved bike path heading west from Boulder Junction. It appeared well used by other riders.

A man and two women zipped by from the other direction; we smiled and waved. I heard him say to his fellow cycling companions, "I'll bet she's in touring heaven!"

Indeed.

STATS
48.75 miles, max speed 37, ride time 4:09, avg speed 11.7, TOTAL 359.11

Seam sealing my rain fly on Main Street in Boulder Junction, oblivious to passersby. A couple I met earlier at the flea market startle me. "You won't meet your schedule doing that," they say. Is my stink eye too obvious? We laugh.

7

A Quarter Short Braggart

Wednesday, June 15, 2016
Lake of the Falls to Hayward, Wisconsin

Awkward in a frilly dress and Easter bonnet, Pat squeezed her mother's hand as they approached the dour Father Hutting. "Happy Easter Father," her mother said.

The priest loomed over the girl. "Father, can I be an altar boy?"

Laughter scoffed from deep within his most holy vestments. "Girls can't be altar boys!"

"Why not?"

He flicked her long braids. "For one thing, you have these."

It took the girl until June to talk her mother into cutting her hair. But this was 1961. Father Hutting would not budge.

Andy and I have routines to help us track the days of the week. For example, Saturday mornings he cooks blueberry pancakes with his homemade sausage, Tuesday is bridge at the senior center, and Wednesday is breakfast at the New Sunrise Café in Lupton. This being my first Wednesday on the road, I'll take his email suggestion and keep our tradition. "You're under budget. You should treat yourself to a meal out." The ACA map shows three restaurants beyond Lake of the Falls.

In 100% humidity, I rustle up quarters, pack, and ride over to the shower building. This time the lights turn on, but I am one quarter short of a shower. Ah, well, at least my phone gets charged. And the seam sealer did its job in last night's storm.

The restaurants are bars, not open for breakfast. I slug a plum and the last of the dried yogurt from home and set to pedaling.

Lovely rolling road. Where's Lou when I need her? There is nothing quite like riding in sync with a tandem partner, rising as one out of the saddles to hammer up and over hills. As I push against my seat back (can't stand on a recumbent) I reminisce. It took miles during an ultra-marathon ride for me to figure out how we coordinated a smooth transition without speaking. This: no matter where Lou's attention was, she sensed my body coil as our right legs powered in phase to 12:00; together we pulled against the handlebars, grabbing leverage with our weight for the downstroke; as I lifted, she lifted. Seamless.

Today I am in sync with myself and I feel great.

Twenty-four joyous miles transport me into Butternut. Passing a farmer getting into his truck at a feed store, I ask, "Is there a good breakfast place in town?"

He points me to Schienebeck's Shanty on the south side. At last, I get my usual—two eggs over medium, burned bacon, hash browns, wheat toast, and coffee. Oh, and the Wednesday special, a potato donut. Never heard of a potato donut and doubt I'll order one again. It is so soggy I wonder if the cook let it cool on an open windowsill last night.

My FitBit tells me I'm burning more than I'm eating, but I can't finish everything. *Guess I'm eating enough.*

The humidity turns into off and on-again rain but nothing severe, in fact, the sun shines in between drops. It is early afternoon when I reach my sixty-mile target for the day. Wisconsin State Highway 77 is smooth with a shoulder and I am strong. Twenty-three more miles to Hayward? *I got this.*

Eight miles later the rain comes harsh.

Dow's Corner Bar materializes in the middle of nowhere, convenient on my side of the road. Without a second thought, I pull under its eave. Parked on the side are two SUVs loaded with canoes and bikes. Inside the dusky bar, the two couples they belong to, wearing cycling shorts and technical clothing perched at a high wood table, are easy to spot. A bear-of-a-local man sits at the wood-topped bar and a grandmotherly woman works behind it.

Greetings hail.

"Do you serve food?" I ask Grandma, who turns out to be Pam, the owner.

"No, just chips or popcorn."

A woman cyclist pipes up, "She has a pretty good cup of hot tea."

Tea sounds better than beer to warm up, so I order it with honey. And popcorn. "May I join you?" I ask these comrades-on-two-wheels.

"Come on over. We're staying at another friend's cabin for the weekend. We hoped to take a ride today, too, but this rain." One man shakes his head.

"Wasn't too bad earlier." I climb onto a stool. "Except for the bear."

All heads turn.

"A couple days ago a woman stops me at the top of a hill and says, 'Be careful, I just saw a bear where you're headed.' Never saw that bear, but this morning I'm cruising along and look up and think I see another cyclist coming my way. Nope. It was a bear. A huge black bear. He was strolling toward me in the other lane."

"What did you do?" The other woman slides to the edge of her stool.

"I stopped. Easy to do going uphill at five miles per hour. He didn't see me. As loud as I could I hollered 'Bear!' He bounced to a stop, got all huffed up, his ears stood up like satellite dishes. We had a standoff."

I take a sip of too-hot tea and almost spit it back into the mug.

"I didn't know what else to do, so I started ringing my little bicycle bell like crazy. It worked. He bounded off into the woods. My heart was beating, I tell ya. 'Yo bear! Didn't Coyote warn you about me?' I yelled that for quite a while."

Everyone laughs.

One would think Pam poured me Long Island Iced Tea the way my tongue loosens. They want to hear about my trip to celebrate B'76, but on I brag. Pacific Atlantic Cycling (PAC) tour in 1988 with Lou on the tandem (San Diego, California to Jacksonville, Florida in seventeen days), our twenty-four-hour women's tandem record (422.5 miles) in 1986, the Paris-Brest-Paris (PBP) women's tandem record (1250 kilometers across France) in 1987.

When I mention the Iditabike (a mountain bike race in Alaska on 200 miles of the Iditarod trail, held two weeks before the infamous dog sled race) in February 1990, the local man asks, "Have you ever done the Birkie?"

"No, I wish I had." Not sure why I never did.

He refers to the American Birkebeiner, since 1973 the world's largest cross-country ski event, over fifty kilometers from Cable to Hayward, Wisconsin. "I did it every year for more than twenty years." His barrel chest puffs a bit wider. "My son has done it too."

His story impresses all of us. Tall tales in Dow's Corner Bar keeps everyone bright, even if the day is not. The rain falls harder.

"So, where you headed today?" The other man of the foursome asks.

"The KOA in Hayward."

"Do you want a ride?" When I choke a chuckle, he adds, "Oh, maybe that's sacrilege or something, taking a ride."

"No, I've got nothing to prove. And I never turn down a ride, but no way you can take my recumbent."

"Oh, my husband will figure it out," the man's wife says.

Before I blink, we are outside shuffling a bike off one rack and into the other vehicle, along with several pieces of luggage. As quick as I dismantle my load, the wife grabs my gear and tosses it into the now-clear back seat.

"I have to take off my fairing," I say, remembering a cyclist on the Dick Alan Lansing to Mackinaw (DALMAC) ride one year. He put his recumbent (with fairing mounted) in the back of his pickup. Somewhere along Interstate 75 heading north the wind caught the fairing and tossed the bike into the ditch. The long piece of Lexan™ plastic, designed for reducing drag in a headwind, also makes an excellent sail. His bike was an unrepairable mess.

Four plastic thumbnuts later and my Zzipper fairing is off, rolled up, and secured with a bungee cord. Sure enough, the husband and I mount my Tour Easy onto his rack. He obliges me; I secure it with his straps and more of my bungees.

"I'll drive careful," he promises as we drive off, rain still pelting. From the back seat, I keep an eye on my precious cargo. The bike remains stable.

The wife says, "He worried what we'd do today. Thanks for giving us this adventure!"

Visibility is terrible, riding in this rain would have been miserable. By the time we rectify a wrong turn near Hayward to find the KOA it eases. Perfect timing. I expect to unload at the camp office, but this couple will have none of it. They wait for me to register and drive me straight to a sandy campsite tucked beneath towering pines.

While I am aghast at the $41 fee (the last time I stayed at a KOA was with

Jim in 1975), I take advantage of the amenities. An outdoor "Kamping Kitchen" is a pavilion with eight cooking stations in a circle, surrounded by picnic tables. Each station has an electric stovetop with four burners, a kitchen sink, and plenty of electrical outlets. I charge my phone, cook dinner, clean up in comfort, and open my journal.

I want to write about logistics: how to decide the next day's destination; how to coordinate food shopping, clothes laundering, phone charging, and shower-taking. I want to write how I feel about the first week of my two-week minimum deadline. In seven days, I've clocked 427 miles. Thirty years ago, Lou and I rode almost this far in twenty-four hours. Incredible to me now.

This long, stimulating day keeps my pen from paper, the best I can do is an email to Lou, thanking her for the experience. I soak in a hot shower and retire, grateful for a dry tent and the generosity of strangers.

STATS
68.41 miles (+ 17 driven to Hayward), max speed 32.5, ride time 5:17, avg speed 12.9, TOTAL 427.54, Averaging 61/day for the first week.

My first haircut at age five.

8

A Week to get Into the Groove

Thursday, June 16, 2016
Hayward to Cumberland, Wisconsin

Two weeks after a tonsillectomy? Scarlet fever. Didn't matter how many popsicles she slurped, the girl thought her throat would never stop throbbing. Finally, it did, and she remembered.

"George!" The red-eared slider named after her great uncle George, who worked on the railroad across the northern Midwest. George, the turtle who turned out to be "Georgette."

The girl ran to the plastic lagoon stored on the bottom shelf of a wire plant table. Dry as a desert. George's likewise dried up feet and head hung lifeless from her shell.

George was her responsibility; still she was heartbroken. "Didn't anyone care for her?"

The girl wrapped George in Kleenex, nestled her in a Maxwell House coffee can, and buried her in a backyard corner.

A fellow camper interrupts my packing. The man and his wife are heading home to Grand Rapids, Michigan, from a two-week road trip to Yellowstone and the Grand Tetons. They've owned a home inspection business for eleven years.

"I can work from the road because our employees are doing a good job for us. Where're you headed?" I squeeze in my spiel. "Montana, eh? We met a cyclist in Montana doing a self-supported race from Seattle to Washington. He broke a shifter cable, so we drove him to a bike store in Boise. He told us he rode fifty-four hours straight and only slept a couple hours here and there. He only carried one extra shirt and hadn't taken a shower. Boy, did he smell bad! You look to be doing better than he was."

What I was not doing was racing across the country. Or riding across in seventeen days as I did on PAC tour (a test which helped me decide against tackling the non-stop RAAM, a supported race unlike the event of the cyclist this man helped). He breathes. I take advantage. "Touring is where my heart is."

He jumps right back in. "I was into bikes when I was a kid. Bought my first good ten-speed bike on a Friday and rode it straight to my friend's house. I said to him, 'Let's ride to the beach.' We took off and rode forty miles to Grand Haven, on Lake Michigan. Riding a bike is freedom."

I nod, eager to get riding. A scene in Hanks' movie plays in my head. School bullies chase Gump as he walks home with his childhood friend, Jenny. She yells, "Run, Forrest, run!" The heavy leg braces he wears to correct his spine hinders his gait. He drives on and the braces disintegrate, freeing him to sprint away.

Ride, Headwind Queen, ride.

Morning fog breaks with sunshine. The ACA route weaves up, down, and around cottages nestled on sparkling lakes. The smooth asphalt bounding through tall hardwood forests would be more fun if my legs had snap. A light tailwind tries to ease my way, but like the princess and the pea, an irritation between my butt cheeks bucks me off my seat after only twenty miles. The tight fabric of my Lycra cycling shorts, compounded by yesterday's rain and this morning's growing heat, is the culprit. A deal-breaker if it gets worse. I need to change.

The good thing about riding a recumbent is that regular baggy shorts work fine. The bad thing about rotating three riding outfits? Only two pairs of baggies.

A cramped restroom in a gas station party store offers private space for swapping shorts and a strategic "gop" of A & D ointment. I cinch the belt on baggies that are getting bigger. Mental note: find a third (and smaller) pair.

Hills increase in distance and steepness, dipping to cross a river or stream. My dead legs must walk. What's that click? A loose or broken spoke? Both wheels are okay. The sound goes away when I pedal, returns when I walk.

Bloody hell.

The hills. Are getting to me. I sing a made-up ditty:

I'm going D O W N
down to the river
and I'm climbing back out

Fewer rivers and more hills create an addition:

there ain't no river!
you're going down, down
there ain't no coming back out!

County Road (CR) F smooths out like how my Swiss Army knife spreads chunky peanut butter on the wheat bread smushed from days in my pannier. Lead legs loosen. Maybe the next thirty-five miles won't be a drudge. Seven miles later the ACA map directs a right turn onto CR DD. Turn off of this beautiful road onto … that steep hill on a sharp curve, with crumbled pavement?

I pull over to consider an alternative. An SUV rolls to a stop and a barefoot young lady exits from the driver's side. "Need help?"

"No, I just don't want to go up that hill." I gesture to CR DD.

"Don't blame ya. You going to Birchwood?"

Off-route, but in the right direction. "Yep."

"Stay on F one mile and turn right onto 48. It'll take you right into Birchwood. Three miles either way and you won't have all those nasty hills."

The words slip out of my mouth, "Thank you, deary!"

I follow her advice, to my great delight, and reminisce about a long-ago day when I diverted off the B'76 route.

The steep Missouri hills and their heat-softened asphalt were more challenging than Rocky Mountain passes. Our last day in the sizzling state professed to be a long eighty miles. Even with a scheduled rest day at the university in Carbondale, Illinois, I was done fighting the Ozarks. And done following ACA routing on back roads with abrupt and sheer climbs, just to avoid a little traffic. Must be the city girl in me. (And Mom's words: "Go play in traffic, Patti!")

Back roads aren't graded like those more traveled. What, did the Road Commission tire of bulldozing knolls into manageable slopes? Or did they run out of money?

Bidding goodbye to my group, I used a state map to plot a saner route along state highways A few miles from Carbondale's welcome sign, a beat-up pickup truck passed and pulled onto the shoulder. An African American couple waved me down.

"Are you with a Bikecentennial group? Are you staying at the college? Here, let's put your bike in back, we'll take you there."

The couple said they helped cyclists all summer. How could I say no? They drove me to the Post Office to collect our mail and treated me to pizza and beer before dropping me at the dorms.

I was the only one of our group to make it to Carbondale that night. The others stopped short and limped into town the next day.

Today I make Cumberland on my own steam, tired as I am, grateful for an easy downhill. The ACA map shows a population of 54,829, with all services, so I look forward to a not-so-tiny grocery store. A hardware store would be nice, too. I'm missing a wing nut from the top right screw of my fairing. When I removed the fairing yesterday, I made a mental note to keep an eye on them. So much for mental notes.

A hardware and Louie's Finer Meats, within spitting distance to the Eagle Point Campground, are not mirages. The hardware closed a half hour ago. No worries, I'm sure there's a nut and bolt in my repair bag.

Louie's is not closed. Tempted by grilled brats and burgers for sale in the parking lot, the prospect of what's inside lures me. I hope I'm not drooling like our black lab, Gus, when he sits on his mat before eating. Yum! An inch-thick pork chop and instant mashed potatoes for me tonight.

Eagle Point is an apt name for the city park that fills a narrow peninsula on Beaver Dam Lake at the end of a suburban neighborhood. Registration is at the camp host's RV parked at the entrance. A gray-haired woman in a flowery cotton top shows me grassy sites under 100-year-old oaks, near the shower and bathroom building, a bit farther from several large RVs.

"Choose wherever you like." She smiles and collects the $15 fee. I am happy to hand it over.

It's only taken eight days of setting up my tent before I learn an easier way. Two long, shock-corded poles slide through fabric sleeves on one diagonal, the third crosses over from the other and connects with plastic clips. Without staking the corners first, the two poles in the sleeves flop down and it's no fun trying to string the third pole. With corners staked, the sleeve poles stand right up, making the last one easy.

See, it does take a week to get into the groove.

Dinner is delicious, shower hot, and a soothing sunset over the water. I have a long conversation with Mom on FaceTime, our daily chat might become one of the best things between us. Mom thinks I should keep my mind on reaching Missoula.

"By then you'll be tired of buying and cooking meals, setting up camp and breaking it down, and you'll just want to come home." She sounds resigned about me doing this, hopeful I'll not keep riding.

Funny, I've been wondering too. Pressure to meet a deadline keeps me from my usual exploring. When Andy and I cycle around home, he rolls his eyes when I say, "I wonder what's down that road?" The idea of a meandering Forrest Gump ride further west before heading home has faded. Will getting to Missoula be enough? Will I be done? A long train ride is something else I've always wanted to do. Could it be my ticket home? If so, I could enjoy a long drive to the east coast with Andy for his niece's wedding.

I miss him terribly.

And the annoying click? An all-day puzzle. With every step pushing uphill, a tube of lip balm in my baggies' right-hand pocket knocked against an aluminum fuel bottle attached to the bike's seat frame.

STATS
69.53 miles, max speed 34, ride time 5:32, avg speed 12.5, TOTAL 497.08

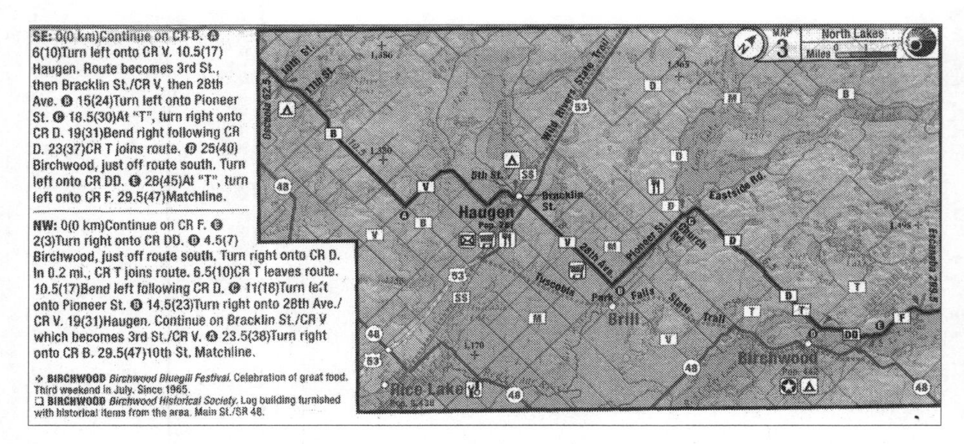

ACA North Lakes Route, Section One, Map Three.
Fewer hills following CR F to State Highway 48, through Birchwood to CR V, and back on the ACA route.

9

Ignoring my own Advice

Friday, June 17, 2016
Cumberland to Interstate State Park, St. Croix Falls, Wisconsin

Her mother found her on St. Martin's Ave. in a snowbank next to a parked car. The girl tossed snow into the air and giggled when it sparkled onto her glistening face.

"Momma!" The girl raised her rosy cheeks in delight.

The mother smiled, took the girl's hand, and walked her the rest of the way to school. It was her first week of kindergarten, back when children started school on the half-year if their birthdays came after January. The early phone call from the office stunned the mother. "No, she should be in school. No, I don't know where she is."

The girl skipped next to her mother until they got to the imposing brick public school. "I don't want to go!"

The mother picked her up, carried her inside, and passed the kicking and screaming girl to the teacher. "Just go," the teacher told the mother.

Tile stung when the teacher threw the girl to the floor. She was forever colored skeptic.

A cacophony of crows pecks at my unconscious. The rising sun blinks my crusty eyelids open. A silhouette parade of grass, clover, and dandelions march at eye level, my cheek stuck to the tent floor.

"Fringe ant."

Franz Kafka? No, one of my earliest memories. My brother's voice. *"That's what you are, Patti."*

Rick knelt next to me as I squatted on the cool cement floor of our basement pantry, shelves stocked with tomatoes, beans, and applesauce

Mom canned. A lone bulb in the ceiling cast just enough light to make the mass of sugar ants swarming in a pile the size of a dinner plate look like one shimmering being.

Mesmerized, I reached out. An ant tickled up my finger.

"See those ants over there?" Rick pointed into the shadows. "Those are fringe ants. They don't stay with the rest. But as they explore, their movements pull the mass of other ants in their direction."

I forgot all about the ant that worked its way onto my arm and turned my attention to the handful of fringe ants scurrying in haphazard patterns. I looked back, the pile had shifted. He was right.

Rick and I recently shared growing-up stories. I asked him what possessed him to declare me a fringe ant. "You couldn't have been nine or ten years old."

He didn't remember the incident and shrugged, "Guess I'm a fringe ant too."

Indeed.

Many a night after I massaged Liz to sleep during her long decline, Rick and I sat in the dark on his front porch and communed over her impending death. Every drag on his cigarette exposed the exhaustion on his face.

"It was okay," he wrote in an essay to me after all was done. "A good exhaustion come from holy work, filled with little rituals of movement, touch, and care." Rituals as a vehicle to share meaning. "Meanings from coincidence, ritual from conscious recognition of repetition, stories woven from reasoned improbability, all so that one's life could be made sense of—something very different from having it make sense."

In his essay, Rick said I was the first person he "helped become a human being." Me, the not-another-sister sibling; me, the tom-boy he dragged along on bike rides and adventures; me, the subject on which he practiced rituals of storytelling. Funny how I always thought he meant "made" me.

"By the way, you didn't 'make' me," I said during our fringe ant discussion. "But I appreciate being the younger brother you didn't get." (Right away, anyway. Two brothers and a sister came later.)

"I advocated for you, you know," Rick said. "I bugged Mom and Dad to get you that rocking horse and let you dress up as a cowboy and Indian."

I didn't know.

Wiggling to reach my camera and capture the Kafkaesque scene on my tent wall, I whisper, "Thanks, Rick."

The road west from Cumberland is bump-ity, bump-ity, bump-ity through familiar Midwest terrain. Acres of farmland roll between hardwood forests. I breathe deep and inhale the negative ions seeping from stands of pines. In a field of luxurious grasses that brush their bellies as they munch, three cows stand cheek to cheek and turn heads in unison.

A few miles beyond the no-services town of Bunyan, I stop at Pap's General Store to empty and refill with Gatorade and a stick of hunter's sausage. Good decision, I didn't realize I missed a turn.

Pap's is a long, low steel-roofed general store squatting behind a flagpole with Old Glory in full splendor. Under a covered porch that runs the length of the building are racks of houseplants, rickety wood benches, a vending machine, and a propane tank cage. The interior is dusty, with uneven wood floors and grime-streaked windows.

The girl who rings up my purchase, poured into skinny jeans with long blond hair as dusty as the store, looks to not even be twenty. But everyone looks younger to me these days. On the customer side of the counter, an even-younger looking dark-haired, pony-tailed girl rolls back and forth on an off-kilter office chair. She wears torn jeans, and a faded red t-shirt of a head-banger band. The two of them are coordinating plans to drive to the dentist together on Monday. Both cracked a tooth this week.

"See?" Ponytail stands up and opens her mouth.

Skinny Jeans rubs her cheek. "I'm just going to have him pull mine."

"Me too, but I don't know how I'll get through the weekend. I can't take any pain meds."

"I'll do the same as last night. Whiskey."

"Hey, that can work," I pipe in. "When my younger brothers and sister cut teeth, my mom dipped her finger in whiskey and rubbed their gums."

"What's the smallest bottle I can get?" Ponytail asks. Skinny Jeans reaches behind the counter and hands her a one-shot plastic bottle.

I sympathize with the rural poor who opt to have a tooth pulled instead of getting it fixed. When Andy and I owned the bike store, we couldn't afford health care, let alone pay for dental work, and went many years without regular checkups. I broke a molar. By then I was working outside the shop and a regular paycheck allowed me to afford a crown to save the tooth.

Skinny Jeans points me back on route. My legs feel more like pistons than yesterday's stovepipes, so a couple miles of detour aren't much to fret over.

What I envisioned: an early start on a short day; lunch at a trendy bar in Centuria next to a trailhead on the Gandy Dancer State Trail; an easy ten miles on the path to Interstate State Park; and rest or a sightseeing walk into St. Croix Falls.

Five miles from Centuria on a smooth, curvy downhill, Balsam Lake comes into view like a mirage. Blue water, blue sky dotted with puffball clouds, a sandy beach with families picnicking in the shade. A bathhouse brings a quick thought, *I could have a picnic lunch here, cool off with a swim, and probably take a shower too!*

Too late, the hill hurls me past.

I push through the village of Balsam Lake to keep to my original plan, but oh, the smells wafting across the Top Spot Tavern and Grill's outdoor patio. I'm salivating. Surely Centuria, with the bike path and proximity to the border between Wisconsin and Minnesota, will be like this place.

Riding the Gandy Dancer requires a trail pass and the ACA map shows I can buy one at the Glass Bar, located a block from the trailhead in a back-alley part of Centuria. Its name must refer to the glass block windows set into the paint-peeled, cement block building. Even with a Harley-Davidson motorcycle parked at the curb, I'm not sure the place is open. I try its rusty screen door. It opens to an even more decrepit steel door.

Inside the murky bar, a burly man with body hair curling at the edges of his leather vest haggles with the bartender over the price of a burger and fries. His gruff voice echoes off the high tin ceiling. The only other patron sits at the bar, a beer bottle sweating a ring on the distressed wood, arguing through her cell phone.

I take a peek at a ketchup stained, cardboard menu at a tall table near the door. Frozen pizza and an assortment of cheese curds are the only offerings besides burgers and fries. Why, oh why, didn't I heed my own bike touring advice? Stop when a place beckons.

With the motorcycle guy appeased, the bartender turns his attention to me.

"Do you have rail trail passes?" I tuck the menu back in its place between a napkin holder and salt and pepper shakers. I'm not eating here.

He blanks, then turns to case out the shelf behind the bar. "Uh, no, we don't have any. You could check at the gas station."

"Thanks." I hightail it out. Screw the pass, I aim straight for the trailhead. Its new, air-conditioned restroom looks out of place at the edge of sad Centuria. A middle-aged couple on hybrid bicycles look to be just leaving. Might a conversation distract me from my vision disappointed? "Where you headed?"

"We rode from the state park. We're camping there and are heading back." Over his shoulder, the man adds, "Have a good ride!" Poof, they're gone.

I'm bonking. Not as much "hitting the wall," like in a running marathon, but more like "hangry." Unable to concentrate and aware I'm making poor decisions, little irritations impede my drive to carry on. I refill my water bottles and dig out some beef jerky and a crumbled fig Newton. It'll have to do.

At Interstate Park, a ranger assigns me the last site in the rustic south campground. She points the location on a map.

I deflate. "Three miles downhill?" No sightseeing for me, I won't be climbing out tonight.

"I could put you in the north campground. Still downhill, but only a mile the other direction. It's full, but we can't turn cyclists away. You'll be in an open field with no picnic table or anything."

I can't decide.

"The showers are in the north campground. There's a hiking trail out of the park there. You could walk to the convenience store on 87."

What to do? For sure I'm bonking. The ranger does her best to help.

"If you want to check out the north site first and decide to stay there, just call me. I'll change your paperwork without you needing to come back."

Outside, I try to engage two men bicyclists who've arrived on shiny road bikes decked out with new touring gear.

"We're doing a fifty-mile overnighter, getting ready for a longer tour next month to visit friends in Chicago." And that's it. I'm left watching the back of their salt-stained jerseys disappear into the station.

Am I that crabby? Ah, well. No decision is the easiest decision to make, I push off for the three-mile cruise to my designated site.

Rocky cliffs above the St. Croix River make for a spectacular view as I coast, trying not to think about tomorrow's climb. Camp at last. As I force the last tent stake into hard ground, the two riders roll into the site across, drop gear, set up tents, and take off. To dinner, no doubt.

Not me. I review my stock: half a package of instant, loaded mashed potatoes, a serving of penne pasta and Parmesan cheese, banana chips, dried cherries, and a dark chocolate cherry Kind bar. Oh, and a bag of Skittles a gas station clerk conned me into buying three days ago. I eat everything.

I'm tired. Am I not eating enough? Maybe I'm not strong enough for these incessant hills, or maybe I need a day completely off. I'm afraid of getting behind, don't want to miss the Missoula party.

With only a 50% charge left on my phone and no place to charge it, conversations with Mom and Andy are short. Andy is tired too, from cutting grass and working on the fence. He lends a sympathetic ear as I unload my day and laughs at my eclectic dinner.

"Reminds me of that dinner you made when we stayed at those cabins on the Appalachian Trail. Never had mashed potatoes and fried pepperoni before."

"We ate it, didn't we? Anything tastes great when you're trashed."

Rain chases me into the tent early. I devise a plan for morning—head for a shower in the north campground (and charge my phone?) and take the shortcut out of the park. If I cross the river into Minnesota through Taylors Falls instead of south at Osceola, I'll cut fourteen miles off the ACA route.

My thought as I lay back to sleep: *I hope I can keep it together.*

STATS

Not an easy "off" day 48.08 miles (three just into my campsite!), max speed 30.5, ride time 4:33, avg speed 10.6, TOTAL 545.17

Rick holds five-day-old me, flanked by Cathy (left) and Sue (right).

Warrior on wooden horse.

10

A Stranger, Better Day

Saturday, June 18, 2016
Interstate State Park, St. Croix Falls to Dalbo, Minnesota

A few years later, this youngster sprawled on her belly in the living room with a Sears and Roebuck Christmas toy catalog creased open to the bicycle section. Elbow deep in blue-green shag carpet, chin cupped in her right palm, the girl's feet swung up at the knees and pedaled air.

Over a picture of a bike she doodled a sleeping bag strapped to the handlebars, bags hung on a rear rack, with a rolled-up tent on top.

She dreamed.

Only one section of the long climb to the north campground demands that I walk. The shower is delectable, and my phone gets charged. The short-cut is easy to find. A trail? No. Cement stairs edged by a narrow curb rise to disappear at the top of a steep ravine. Can steps be deep and shallow at once? These are. The depth from front to back is at least two feet, the height less than six inches.

Ah, well, what's a few steps to an old stair-climber like me?

Except for these stairs being cement, the setting reminds me of the 190-some open-backed, wood steps at Bloomer Park in Rochester Hills, Michigan. The start of my stair-climbing career.

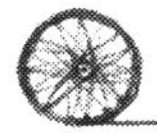

"Let's pack our running shoes and ride out to Bloomer." It was the early 1980s. Twelve tandem-riding miles later, Lou had me changing shoes for a new workout. "We'll do ten sets." Ten times down and up those bloody stairs, I'm not sure how I made it home afterwards. Brutal, but fantastic for building leg (and heart) strength, a regular dose of stairs produced longer times standing out of the saddle on the bike.

It didn't end there. At the annual Thanksgiving Day Turkey Trot 10K footrace in Detroit, a fellow cyclist provoked us. "You two should apply to do the Run Up the Empire State Building." Yep, apply. With résumés. The winter of 1988 found us racing a vertical fifth of a mile, eighty-six flights up to the observation deck of the iconic New York building. By 1990, stairclimb racing made it to Detroit. In the belly of the city's Renaissance Center (RenCen), I won first place women in the one-time-up Race to the Summit, then turned around to race the Vertical Mile Marathon, eight times up seventy-two floors. Several shorter stairclimbs around the city, followed by a run up Toronto's CNN tower, and still no end.

Oh, I could blame Lou for the craziness. But noooo…this was all my idea. "No one has ever climbed stairs for twenty-four hours." And so, I did.

Surely these forested stairs will be easier than the 68,992 steps (12,276 more vertical feet than Mount Everest) I managed at the Ren Cen in February 1991. Then again, I didn't have a loaded touring bicycle to drag up too.

Maybe I can roll my rig on the curb. I set the front wheel on the edge and wrangle the weighted rear wheel up. Left hand gripping the handlebar above me, I shoulder the seat back and dig in. The front end skips off into the dirt. The slope is too sharp, the curb too thin, I cannot keep both wheels engaged.

Okay. I lift the front wheel onto the third step, roll the bike until the rear wheel hits the first step, then lift the rear wheel onto the second step. Raise the front wheel to the fourth and repeat. Cumbersome, but doable. Forty-eight times.

Now the steps are less deep, and the bike is stuck. Sweat burns my eyes, retreat not an option. I muscle to a landing, remove gear, and Sherpa everything thirty-eight more steps to the top. At last, here a trail leads to the highway. Three trips for gear and a fourth to carry up the bike. My shower is wasted.

Stair travail over, smooth country roads and light traffic are a joy. Llamas graze under trees in a lush pasture and turn their heads as I glide by. Never passing an opportunity, I take advantage of a porta-potty behind a manicured ball field in Sunrise, Minnesota, the proud birthplace of actor Richard Widmark.

I got this.

The ACA map designates Donn Olson's farm, west of Dalbo, a Cyclists' Only Camping/Lodging. I call and leave a message. No worries, riders can pitch a tent in the yard even if the Olson's aren't home.

Dalbo, population eighty, is nothing more than a small post office across from a bar. The post office is closed, but the cement porch at the entry is shady, so I get off the bike to give the Olson's another call.

Donn answers. "Oh, you're less than a mile from us. You'll be here in no time."

An American flag on a pole at the edge of Olson's wide asphalt driveway waves me in like a beacon. A white farmhouse wrapped with a porch is barely visible through old-growth pines and hardwoods. Amongst miles of crop fields, the homestead is an oasis.

Leaning against a tree is a piece of plywood painted white with black letters and a bicycle. "Adventure Bicyclists Bunkhouse." A pale-yellow, steel-sided barn nestles behind pines that are taller than the old silo squatting next to it. Another sign on an overhang support confirms this is my refuge for the night. Before I dismount, Donn strides from the house.

"Welcome!" He holds his hand out for a shake. The smile lighting up his face says everything about him. "Ah, a recumbent! A guy left here this morning going west. He was on a three-wheeled recumbent."

Donn wants to know everything about my trip, but first, he opens the door to the bunkhouse.

"Roll your bike right inside. You have the place to yourself tonight."

Inside is what I'd expect a converted old barn to look like—rough-hewn wood walls, log ceiling supports, and a cement floor—minus the dust and animal odors. The narrow main room runs almost the length of the building; a round kitchen table in the middle has five mismatched office chairs waiting for sitters. At the far end a flat screen television on a bureau faces a cushy couch. In the window is an air conditioner.

Four rooms line the wall opposite the entry, each with heavy wood doors and handmade latches, and wood bunks with exercise mats for mattresses (each with a pillow kept clean in a garbage bag). To the right is a kitchen equipped with two refrigerators, a hot water tank, sink, microwave, hotplate, pizza oven, and coffee pot. Plastic bins stored on shelves under the counters contain pots and pans, utensils, paper plates, and small boxes of cereal, breakfast bars, cookies, chips, nuts, and other snacks. Suggested donations marked on these (and canned goods stacked next to the refrigerator) cannot cover their cost. Frozen pizzas are two dollars. Pop, Gatorade, juice, and chocolate candy bars are one dollar. Bread, cheese, eggs, and milk are free for the taking.

Cost to stay? Free.

"I just ask that you sign this." Donn picks up a small black notebook held closed with a rubber band. The year 2016 is inked in white on the cover. Earlier years' books are piled on a shelf under a sign with instructions and rules of courtesy.

Behind the kitchen, a wooden staircase leads to a loft filled with enough cots to accommodate a large group. The outhouse is weathered barn wood, complete with a crescent moon cut in the door. Next to it is the outdoor shower made of the same gray wood.

"I used to heat the shower with the sun, but this year I put in the water heater," Donn says. "We have wireless Internet, but to use your cell phone you'll need to go out outside." The steel roof and siding prevent a signal, even though there is a cell tower stationed in the middle of his 100 acres. "I rent out the land to other farmers these days, but my best profit comes from the cell tower." He brags about his negotiating skills with the telecommunication companies. "They offered me $500 a month, but I checked around and other farmers earned between $500 and $1000. So, I held firm for $750."

Given the well-ordered details in the Bunkhouse, it is no surprise that Donn is retired military.

He leaves me to settle in, promising to return to chat. "Some like to say they've slept in a silo. You can stay out there if you like." I exit with him to check out the clean and dry silo. Two cots sit waiting, surrounded by photos of visiting cyclists on the wall. I'm sure I'm not the first touring cyclist to think I've entered bunkhouse heaven.

The pictures, and the two corkboards in the bunkhouse highlighting this year's cyclists, newspaper clippings, and a world map, throws me into a nostalgic mood. In 1976, our group, designated as "camping," set up our tents in campgrounds, city parks, or schoolyards. B'76 "bike-inn" groups paid more for indoor accommodations—bed-and-breakfasts, churches, and even tee-pees. Every few weeks we stayed in a bike-inn. When we didn't, we still stopped to read messages from other riders pinned on bulletin boards, notes identifying the best places to get ice cream, homemade pies, or all-you-can-eat buffets.

Why did it never dawn on me I might meet other riders on this trip?

On my first trip with Robin, no one we met had ever seen a bicycle tourist. Riding with my brother in 1975 was much different. "Oh, a couple rode through here last week that started in New York," a park ranger might say as we checked in. "We've had cyclists from all over!"

In sheer numbers alone, baby boomers pushed the bicycle boom of the early 1970s. (I like to think we also bucked gas-guzzling cars for a more planet-friendly mode of transportation.) The ten-speed boom was short-lived and interest in mountain bikes grew in the 1980s, but during the eleven years we operated our bike store, only a handful of our customers geared up for an extensive tour.

Guess I've been away from the open road too long.

Donn returns with a cup of coffee and plants himself in one of the office chairs.

"I never knew we were on the Northern Tier Route. Heck, I never even knew about bicycle tourists. One day in 2005, two cyclists knocked on our door. A storm was brewing, and they were exhausted. They asked if they could stay in my barn." He welcomed them and learned about the ACA route that went past his door. "I contacted the ACA and told them I'm a bicycle-friendly farmer. Not many places to camp around here, so I asked them to let riders know they can stay here."

Donn converted one of his barns into the bunkhouse and each year added improvements like the water heater and food choices. Over 200 cyclists visited one busy summer. "My wife likes to plant things. This is my hobby. I would never ride across country like you do, but this gives me a chance to live through you. Besides, it feels good to help people. And I'm the one that got the ACA to add the rail-trail alternate route. They had riders going all the way north to Bemidji. Why do that when we have these excellent trails?"

"That's the route I'm taking," I say. Over 100 miles of paved rail-trail sounds great to me."

Donn takes his leave. I splurge on a supreme pizza (and eat the whole thing) and compose a note for Donn's little black book. I'm the fifteenth rider to visit this summer of 2016. After the required "name, rank, and serial number," I add a special thank you.

"Your bunkhouse came at a good time; yesterday was a struggle and I was very discouraged. But finding this place and you Donn, and your wife, brings back the memories from '76 and what made the ride so special to me, a young punk at age twenty. I was uplifted by the generous and welcoming people along the way who seemed to be caught up in our adventure."

An incessant pounding on the kitchen side door breaks my concentration. Before I can get up, the door squeaks followed by a man's voice.

"Hello! Anyone here?"

I round the corner to find a thirty-something man dressed in cut-off jeans, a soiled tank top, and untied tennis shoes. His eyes dart around the room. "Hey, have you seen anyone else here? I'm looking for someone. He isn't wearing shoes. Has he been here?"

"There's no one else here."

He shifts from one foot to the other.

"Who are you looking for? Your son?"

He backs out. "Never mind. Something just happened to me. I think I need to sleep on it."

And leaves. Just like that, he speeds away in a silver Silverado. Less than an hour later he's back. A woman exits the passenger side and walks around the truck. She's barefoot.

"I know this sounds weird," the man begins. "But have you seen a man with a huge black beard and no shoes?"

"Nope. Haven't seen anyone."

He holds himself, arms folded halfway around his torso. The woman is his wife. This afternoon while he was cutting grass and she sat on their porch in Dalbo, a stranger walked up the road carrying a wooden paddle.

"He had the blackest feet I ever saw. I couldn't get over them. I couldn't stop looking at them. And a huge beard. I asked if he needed a drink of water. He said no. I asked if he needed help. He said no. I asked where he was going. He said he was walking from Anoka to Wakhon. That's a long way! I know because we have a cabin up there. I asked him why he was going there, and he said, 'I'm going to get my paddle varnished.' That's when I noticed there were all these weird markings carved into it."

His wife adds, "And the weirdest thing was my two dogs. They bark at everyone, but they just came over and hung out by this guy and didn't bark at all."

"I couldn't stop thinking about him. After he walked away, I decided I needed to find him. But he wasn't anywhere! He couldn't have walked very far. That's when I came here before, I know that bikers stay here, and I wondered if he came here. I drove up and down all these roads and there's no sign of him, it's like he disappeared!"

His wife goes on about their dogs and how calm they were by the stranger. The man goes on about the stranger's black feet. "I am not a religious man, but I have to sleep on this. I shook his hand! I can't figure out where he went." He gazes over the pines. "You're going to think I'm crazy, but I think he was Jesus." He bows his head and shakes it. "I need to sleep on this, I'm not a church-goer. But it's the only thing that makes sense."

I put my hand on his shoulder. "You're not crazy. Obviously, something happened to you."

Just then Donn and his wife drive up in their car and join us. The man gives a shortened version of his strange story. Donn stifles a laugh about the paddle getting varnished. I don't get the connotation at first, I guess I'm taken by the young man's intensity. Sometimes I feel a bit like Lucy from the Peanuts cartoon, with her "Psychiatric Help 5 Cents" sign above her head. I'm not a psychiatrist, but I must have an open face or a knack for listening, how strangers so often share their inner thoughts.

The couple leaves to continue their search for the man with black feet carrying a paddle. Donn says, "Do me a favor. Keep your bunk door closed and latched tonight. I've never had a problem with cyclists staying here, but that was crazy."

Not sure if he's worried the man from Dalbo will come back or if the black-footed stranger might appear. It feels silly, but I do as Donn says.

STATS
55.79 miles, max speed 26.5, ride time 4:40, avg speed 11.9, TOTAL 600.98

Me and Donn at the Bunkhouse.

11

With a Little Help

Sunday, June 19, 2016
Dalbo to Bowlus, Minnesota

Pat's older-by-seven-years brother rolled his new English racer ten-speed bicycle up the driveway. Rick saved his paper route money to buy it. The click, click, click music of the freewheel enchanted her. In one leap, she cleared three steps to the green, wooden side door. Metal dust tickled her nose as she pressed her face against the screen, watching him wheel the bike into the garage.

Dreams turned to lust. The girl pulled herself away, the side of her face grid marked like a map leading out of the city. She leaned against the knotty-pine wall to catch her breath.

Later, she snuck into the garage, stepped on an overturned bucket, and climbed onto the not-yet-broken-in leather saddle. Perched atop the spindly frame, feet dangling and a long reach to the dropped handlebars, the wall in front of her was not laced with spider webs and water stains. No, stretching out ahead was miles of smooth roads, sunlight and shade shifting like the shutter of a home-movie.

No strange incidents all night. Every morning brings a new day, fresh with opportunities. Today? Scrambled farm eggs, toast, and brewed coffee. *Thanks Donn.* On the road before seven, I am eager and hopeful in a not-too-stiff breeze. It's going to be hot.

For at least a quarter mile, six goldfinches flit about me, land on the wire gracing the side of the quiet country road, then fly ahead like an escort. Many mornings I notice birds standing guard. Red-winged blackbirds voice "turk, turk" or "chit, chit," depending on the day. Is one a good morning, the other a warning? Either way I reply, "Good morning birdies!"

The eighteen miles northwest to Milaka is a south-wind-wonderful, more-than-fourteen-miles-per-hour start. When I change into clean clothes at a laundry mat, I grab my tube of A & D ointment. Still a bit of chafing. *Don't forget to put it back. Wait, is that my mother's voice?* "Back" is into a wallet-sized bag velcroed to one of the bungee cords. It's filled with stuff I might need through the day, like lip balm (a solution to that annoying click) and sunblock.

Ten miles west of Milaka, the empty road eases north and I fly past cornfields. What a difference a friendly wind makes.

Wait. Did something drop? I glance back. Nothing. A second later I am sure something else drops. There, in the center of the road, sure enough, my sun block. I forgot to zip the bag shut. I peer further. *Is that the A & D?* Yes!

Heat rises. Where the road ends at a "T," and the route turns west toward Bowlus, a lonely corner bar beckons. I don't make the same mistake as when I bypassed Balsam Lake. I am the Rum Shack's only customer. Instead of beer, I treat myself to Gatorade, and a bag of chips.

On the road again, the side wind batters average speed. An eagle soars ahead with an occasional wing flap. Me? I am no longer flying. Ah, well, the wind and I duke it out for two miles until the road curves north. Yeehaw, a tailwind. It's only for a couple miles, but I am now the eagle, lost in reverie.

Wait. Another curve. The map showed a straight shot to my next turn. *Did I miss it?*

A group of people sit in the shade on a farmhouse lawn; I bounce down their gravel drive for directions. "Nature Road? I'm not sure where that is," a man says. "Where're you going?"

"Bowlus today." Funny how so many people aren't sure of local road names, they just know where to go.

"To Jordie's?"

"Yes." Jordie's Trail Side Café is across the street from the city park where I hope to camp. The ACA map shows a $10 fee payable at the café.

"Ah, then you'll have to go back at least two miles. It'll be the first paved road on your right. Say hello to Jordie when you get there, she's my cousin."

Ugh. Good thing I stopped before ending up in the next county, it takes more than twice as long to retrace two miles to the missed turn than it did riding four with the tailwind. Now on a straight shot west, my bike lists so low against the cross/head wind I worry about wearing my tires out on the left side. Wrestling my bike against gusts is like wrangling a side-twisting bronc. Glad there isn't much traffic.

What is that? Ahead, a white, oblong thing hovers high above, drifting right to left. *A plane?* Too slow. I keep glancing up as I fight to stay straight. It disappears. There it is. Gone. Back again, still gliding along. I can't figure it out. *Am I losing my mind?*

As if distracting me on purpose, a red-tailed hawk swoops across the road, alights on a tree branch and gazes down. A crow dives in, attacking. The hawk takes flight only to land in another tree and watch me roll by. I forget all about the oblong UFO; my mind takes flight to the day of Dad's funeral luncheon. My sister Cathy and I both ducked as a hawk darted low across the restaurant parking lot.

"Dad told me he'd come back as a bird," Cathy said, "and poop on my head!"

No poop, not him? I yell anyway, "Hi Dad! Happy Father's Day!" I can't say I believe the dead signal the living, or come in signs like visiting cardinals, but since Dad passed, it is uncanny how often red-tailed hawks show themselves. Sure, hanging out along freeways hunting for critters might be normal, but almost every time I drove the 170 miles to the city to visit Mom after he was gone, I'd see three or four of them. Even Andy agreed it was weird.

I scan the sky. No sign of the white thing. Ah, well, the same as much in life, it's a mystery. Keep pedaling.

On an open stretch between Ramey and Little Rock, a bike tourist greets me from the other direction. We stop to chat; I am happy for the break. My forearms hurt from fighting to alternately keep on the road and then out of the oncoming lane.

Bob is from Akron, Ohio. "I left Montana on May 29 and have just over 700 miles to get home. My wife wouldn't let me go alone, so I started out with SAG drivers and a few riders. They didn't make it." (SAG stands for "support and gear.") He hands me his card. As I get off my seat to dig for my wallet, the wind lifts my fairing and almost tosses my bike into the ditch. I lay it down.

"I've been fighting winds for ten days," Bob says. "It once took me over an hour and a half to ride sixteen miles!"

Ha, I'd be happy to travel that fast, this Headwind Queen can't remember gusts so demonic. After leaving Bob, the less than eight miles to Little Rock feels like sixteen. I ache for another breather, but there's nothing here except two houses. One has a shady tree, at least. I lean my bike against its broad trunk and the wind grabs it by the fairing; I leap to catch it before it rolls away without me. Time to take action. Off comes the fairing to get wrapped around my sleeping bag. Phew, I plop my rear end on the grass to rest.

No fairing makes holding a line easier back on the road. Or maybe it's thoughts of hawks and Dad.

When I was eleven, tomboy that I was, a group of Harlow kids and I played softball on the sidewalk. After I underhanded a pitch, I put my mitt in front of my face to shield my eyes from the sun. At the crack of the bat, I dropped my mitt to a total softball eclipse and a second crack—my nose.

The Detroit detective who lived in the house where we played leapt from his porch. "I've had my nose broken lots of times. I can fix it." He snatched my nose and cracked it back.

The adage is true, I saw stars.

When Mom took me to the doctor, my face was so swollen there was nothing he could do. One day, with bruises lighting up my face like a beaten boxer, I followed Dad around the grocery store. A clerk pulled me aside and whispered, "Did your dad do this to you?"

I shrugged the stranger off. My dad? Who helped me regain confidence by playing catch with me? The gentlest man I ever knew, who gave us kids space to find our way and who took pride in the individuals we became? He could never do anything like that, despite the occasional spankings I earned. "I'm sorry, Dad!" I'd cry as he took me aside for a whacking.

"Don't be sorry," he'd say. "Just don't do it again."

I always took him at his word.

The afternoon before Dad passed, most of my siblings and a few nieces and nephews came to see him. The days leading up to his death were not easy to witness. "Terminal agitation," was how the hospice nurse described his sometimes-violent jerking and cries of "help me!"

I became the drug dispenser: Ativan for agitation and morphine for pain. At first, they allowed for periods of calm where he drew back into himself. He looked at each of us as if saying goodbye. My heart burned.

"Whatchya got in there?" Dad pointed to my niece's soccer ball belly. She was six weeks away from her first child, Mom and Dad's tenth great grandkid.

"I don't know." Future-momma rubbed her tummy. "A baby."

"I know what it is, but I'm not telling." His eyes glinted.

His eyes drove to me. He reached out toward my nose. Instantly, I was a child again, him pinching my nose and pulling his hand away, my nose in between his first two fingers. That afternoon Dad made the pinching motion again. As I reached for my nose, the convulsions returned. While his body was not yet dead, his eyes no longer revealed the dad I knew. I bawled.

My brother, Jim, and his wife and daughter, and my younger sister and her girls are having dinner with Mom today. Our first Father's Day without him. I hoped to make it to Bowlus before a FaceTime call, but the wind has other ideas. In Royalton I stop at a busy corner gas station. I can't hear over the wind and traffic, so I text Jim about their schedule. Looks like I might make the six miles to Bowlus in time to catch them together.

Bonkers again, I wander around the station's Holiday Market. Surprise, a chicken Caesar salad. I'm a ravenous lab with a bowl full of kibble. Salad! Seriously, it's the best I've ever eaten.

Fortified, I push on, crossing the not-so-mighty-here, not-worth-picture-taking Mississippi River, and make it to Bowlus in time for the call. My sister-in-law and niece pass the phone around. Mom seems lost sitting with the chatty group. I get little time with her and wish I could give her a hug.

"Here's the deal," Jordie says when I ask about the camping fee. "Come with me." She heads out the front door of her Trail Side Café. The restaurant resides in a century-old building, three cement steps from the sidewalk, and an antique bicycle hangs from an iron pole above its arched doorway.

Jordie points to the pavilion and building in the park, the trailhead for the Soo Line Trail. "For $10 you can set up over there. Or, for nothing, you can stay back here." She leads me through a luscious flower garden and over a small bridge spanning a koi pond. "You can set up under my pavilion, but wait until after 7:00 p.m. or so, when most everyone is done with dinner. You'll have to walk across the street to use the park bathrooms. There's no shower, but in the evening they don't get much use. You could wash up in the sink."

A no-brainer. For the free campsite I splurge at Jordie's Father's Day buffet in honor of Dad. Fried chicken, mashed potatoes, green beans, macaroni and lettuce salads, and dinner rolls. For the first time my calorie intake on my FitBit app is "in the zone" instead of under.

With my tent set up on the brick-paved floor of Jordie's pavilion, it is time to get rid of my sweat and stink at the sink in the park. In the mirror I spot a tick embedded between ribs on my right side. *My ribs are showing!* I pinch it out. The head stays intact with its not-yet-swollen body; its legs wiggle down the drain. Inspecting the tiny puncture in my skin I recall reading if you remove a tick within twenty-four hours, you are unlikely to contract Lyme disease. This was the first day I didn't use bug spray. *Must have picked it up when I sat on the grass where I took my fairing off.*

Not very long ago, I leapt out of my skin when a tick crawled on my neck. Who is this new me that doesn't worry about a tick bite?

Just as I make it back to the shelter, a wild hailstorm hits. The pounding on Jordie's steel roof is deafening. What's not to enjoy about a rollicking rain when I am clean, dry, and stuffed? When the storm breaks, calm brings humidity. I write in my notebook, "I made it feeling better than I ought to, given how I pushed. Maybe the breaks made a difference. Maybe eating more."

I pause, and add, "I think Dad helped."

STATS
68.84 miles, max speed 24.5, ride time 5:42, avg 12.0, TOTAL 659.89

Dad and my niece, Cori, as a baby.

12

Rail Trails and Trains

Monday, June 20, 2016
Bowlus to Chippewa County Park, Brandon, Minnesota

One moment, the girl perched on the handlebars of Rick's paper-route bike, eyes closed against the late summer sun. The next, her chin hit pavement. She opened her eyes—she was under a car, staring at the curb.

"Jeezuz Keerist!" a man yelled. The girl turned her head. Shiny black shoes tapped at her feet.

"You okay, Patti?" Rick reached for her arm. She squirmed out, careful not to bump her head.

"What the hell! I didn't even see you coming." The stranger grabbed her other arm a little too hard. "I opened my door to get out, and you ran right into it, you're lucky the window was down or you woulda' smashed right through it. What are you doing anyway, riding down Pembroke like that? Are you trying to get yourselves killed?"

"I'm okay." The girl shrugged free to brush bits of gravel from old scabs on her knees. Tiny red droplets oozed bright against white scrapes on her tanned skin.

The man pulled a handkerchief from the breast pocket of his suit, rubbed a smudge of dirt off the inside of his door, and slammed it shut.

"We're sorry, mister," Rick said. "Is your car okay?"

"Yeah, yeah, it's alright." He wiped his hands with the handkerchief, shook it, folded it, and put it back into his pocket.

"Rick, your bike!" It was in a heap in the street.

Rick righted it and rolled it a few feet. The front wheel wobbled. "It's okay to get home. You alright to ride again?" She nodded.

"Next time you kids look out!"

"Don't tell Mom or she'll never let you go with me again." Rick whispered in the girl's ear.

In an email, Andy says he misses me. In our morning phone call, he doesn't say it out loud. "Two more days 'til your two-week mark." I visualize him with keys in one hand and phone in the other, ready to rescue with the words, "I'm done."

He knows I won't call an SOS to his ERD, but I confide in him that this ride to Montana is not the Forrest Gump turn-where-the-road-entices, sleep-when-I'm-tired, eat-when-I'm-hungry ride of adolescent dreams. "I feel like I'm on a mission. Gotta get to Missoula on time."

He fishes, "And after that?"

"We'll see." Will I ride further, as warned, or home? A train whistles in my mind, the rhythmic roll of wheels on iron rails in beat with "Ramon." Or another way home?

I miss him too, but only say, "I love you."

The Soo Line Trail leaves Bowlus and turns into the Lake Wobegon Spur. What a joy to cruise on railroad grades in shade and no traffic. A few downed limbs create temporary obstacles, but I manage. Seven miles in, I take an early break in Holdingford at a sparkling new restroom building. Its roof extends over picnic tables on a cement pad surrounded by not yet bloomed flowers planted in brilliant yellow, blue, and red truck tires. Next to the trail sits an old boxcar with murals painted on each side, depicting life in Holdingford during railroad heydays.

I FaceTime Mom. She slouches in the smaller LazyBoy chair where Dad's motorized chair used to be, their jeweler's clock ticking away on the wall behind her. Her voice is tired. "You know, I've only seen your sister Annie once since Dad died."

"Mom, she's been over more than that. I've been with you when she's come to visit."

"Whatever."

Ah, well, it is her reality. She's lonely. Father's Day without Dad yesterday couldn't have been easy.

"Look at this." I turn the phone's camera to a mesmerizing scene. The morning sun catches floating tree fuzzies glowing white against the deep green forest beyond the trail's edge, like fireflies dancing in moonlight.

We talk about my ride, how I'm working hard to make the miles to Missoula. She says, "Think about what it's doing to your body. You'll pay for that the rest of your life!"

To myself I snark, *I wish I had a nickel for every time I heard those words.* I am not quick enough to respond that I am more than pleased what this ride is doing for my body. I sleep well. I love the wiry tightness in my legs as I roll over in my sleeping bag on the hard ground. While often heavy in the morning, these old lugs make my projected mileage every day. And surprise—no knee pain or breathing issues.

I breech the subject of my return. "Well, I am thinking about taking the train home. Another thing I could cross off my 'bucket list.'"

"That's a good idea," she says.

The Lake Wobegon Spur shifts northwest at Albany to become the Lake Wobegon Trail. Like how road conditions change at the county line, fresh and fabulous pavement turns cracked and crumbling. No more bicycle murals, sculptures, or full-service trailheads.

In Freeport, a local rider scrapes the rear wheel of his Rans long wheelbase recumbent so furiously he doesn't notice me. I wonder what he rode through but am disinclined to stop. Before long, headlights from an orange work truck on the trail ahead and the bitter tang of hot tar gives a hint. One young man drives at creeping speed while two others walk behind with a suspended, swinging hose and fill cracks.

"How far have you spread this?" I ask.

"About a mile."

No way to avoid the haphazard streams of stickiness. The memory of a similar problem during B'76 brings a smile. On a road somewhere in Kansas (maybe Missouri?), the tandem couple and I couldn't keep ahead of a work crew spraying tar and dumping pea gravel. Forced to ride behind, we stayed as far right as possible and still kicked tar and pebbles onto our bikes. What a black, gooey mess we had to clean off that evening.

I do not want a repeat, so I take my recumbent for a walk, working my way through tufts of dry grass. A woman cyclist passes me with a look like she thinks I'm crazy. I say, "I don't want wet tar all over my bike!"

At a road crossing she stops to check her hybrid. She must decide it's fine and hops back on. After a half mile I hop back aboard too. Not for long. Here's a second crew.

"We've done two miles," a worker says, "but you should be alright in about 200 feet."

I don't believe him and jump out to the parallel road. A mile or so later the tar is rideable.

Now the Central Lakes State Trail, the route slips by littered backstreets of Osakis, population 1740. I duck off to find a bathroom, get circumvented by construction, and find open doors at the Osakis Community Center. My cycling shoes click-clack on linoleum while I cross a gymnasium-sized room to the restrooms. As I enter, a grandmotherly woman at the sink jumps, one hand reaching for the wall, the other gripping her chest.

"Oh my god! I didn't hear you come in!"

"I'm so sorry. I'm riding my bike to Montana and I needed a restroom. The front doors were open, so I let myself in."

If the bathroom light wasn't already on, I could have sworn her attitude change brightened the room like the flip of a switch. At ease and welcoming she says, "I just finished cleaning things up from the senior lunch and was getting ready to leave."

We get to chatting. "What will you do once you get to Missoula?" she asks.

"My plan was to ride home too, but I'm thinking about taking the train back."

"Oh, my husband loved taking the train." She stares over my shoulder as if looking into the past. "I lost him two years ago. He collected old cars and took the train to get them. He traveled all over the country."

I listen to stories of this woman's lost love, reminded of how much Mom must miss Dad. As I take my leave and apologize for startling her, the woman says, "I'm just glad I was still here for you. I'll pray for you!"

Meeting cyclists on a bicycle path should not be a surprise. Tony is riding from Montana to Gaylord, Michigan and Ron is riding from Montana to Pennsylvania. A retired couple are taking a short cruise on a Greenspeed recumbent trike built for two; they rode west to east across the country in 2002. "We've ridden RAGBRAI many years."

The Des Moines Register newspaper organized RAGBRAI, the "oldest, longest, and largest week-long bicycle touring event in the world," in 1973. The popular "Register's Annual Great Bicycle Ride Across Iowa," now attracts more than 10,000 riders. Myself, I've never ridden RAGBRAI. A crowd that size predicts long lines at restrooms and meals, and potential hazards on the road. I prefer smaller groups or riding alone.

Still, these short breaks to chat with fellow cyclists do wonders for my stamina into the wind today, and the easy grades are a good reprieve from hills.

Chippewa County Park is three miles of cornfields north of Brandon, by way of a road with a nice shoulder and little traffic. The fifteen-site campground rests low between Devils Lake and Little Chippewa Lake. Parked under trees at one end of the rolling grounds is a lone camper trailer. I self-register and pick a site under a gnarly oak tree across from the shower building and pavilion, a sandy beach and fishing pier a short walk away.

Funny how strenuous activity can lead to craving. I sauté Brussel sprouts and a red pepper in olive oil. *There is no better dinner than this.*

STATS
78.12 miles, max speed 23.5, ride time 6:53, avg speed 11.3, TOTAL 748.14

Lori and Len inspect the tar debris on their Gitane tandem, my bicycle at rest in front of them.

13

Shifting to Life on the Road

Tuesday, June 21, 2016
Chippewa County Park, Brandon to Pelican Rapids, Minnesota

The girl's grandmother (her father's mother) had a heart condition requiring restrictions. No stairs or extreme upper body movement. Everyone knew she climbed the stairs to her attic when no one was looking.

This day, the grandmother pulled the dark-stained mahogany stool from against the upright piano. The girl wasn't sure if the keyboard cover squeaked as the grandmother opened it, or if the old woman creaked when she settled herself on the stool.

For a heartbeat, all was silent. The grandmother stretched her arms like a tundra swan shaking its wings free of water, her gnarled fingers lit on the ivories. Fingers danced, arms pumped, body swayed to a marvelous melody.

The girl visioned her grandmother young and lithe. Exquisite possibilities took her breath away. The girl sobbed out of control, ran to the bathroom, and locked herself in.

Light bristles on ripples, tickling reeds. I set water to boil for coffee and oatmeal and take a stroll to Little Chippewa Lake with my Nikon. Blinded through the viewfinder, I consider the scene. Does the sun see itself in the water mirror and, like Snow White's evil queen, ask if she is fair? Does she need her reflection to verify her essence?

I expect not.

And me? Seeing the Chippewa Park sign at the entrance yesterday triggered memories of my volunteer role with Leader Dogs for the Blind (LDB) and the Chippewa Correctional Facility in Kincheloe, Michigan.

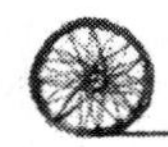

In 2002, LDB started a puppy raising initiative in the North Central Correctional Facility in Rockwell City, Iowa. When its warden transferred to the Fort Dodge Correctional Facility, he took the initiative with him. Hundreds of puppies later, the collaboration between the Iowa Department of Corrections and LDB won the Mutual of America Foundation's 2013 Community Partnership Award. The award came in part because of expansion into Minnesota, Wisconsin, and Michigan.

The initiative proved a win-win for all parties. Puppies raised in prisons graduated as working guide dogs at a high percentage rate. Morale grew and tension fell at the correctional facilities. Recidivism rates for parolees in the program dropped from the national rate of about 50% to 17%. Inmate raisers, proud to give back to society, learned patience, compassion, and gained self-esteem.

LDB brought puppies to the Chippewa Correctional Facility in August 2013 and the volunteer "puppy counselor" responsibilities fell to my friend, Tammy Bartz. I came along to help. The Michigan Department of Corrections gave me permission to bring my camera into the level one facility to document the effort.

In January 2014, Tammy (of the Subway card) and I, with LDB Puppy Development Supervisor Deb Donnelly, introduced the first two Future Leader Dog (FLD) puppies to the Baraga Correctional Facility at the far west side of the UP. More than twenty prison employees and officials applauded our arrival, complete with local news coverage. Oh wait. The applaud wasn't for us, the adorable labs, Axel and Bear, stole the show.

The red-carpet treatment was over, and it was time to deliver the pups to their new raisers. Behind us, the door clanged shut and locked with a shocking note of authority. A buzzer blared a second door open. We were inside, filing through a narrow passageway jammed with inmates. A louder buzzer echoed from speakers in the ceiling, inner doors swung wide, more men mingled. Our progress stalled at an open intersection of hallways. My neck pimpled.

"What kind of camera is that?" A grizzled man in a blue and orange prison uniform strained over my shoulder, hands clasped behind his back as if taking an afternoon stroll.

"A Nikon." I turned away to capture a shot of FLD Bear. The nine-week-old black Lab pulled against his leash toward a not-so-private restroom.

The inmate shadowed me. "You know, you can point your flash to the ceiling and get more light that way."

"Yes." He wasn't telling me anything I didn't already know.

Another inmate bumped my arm. I stole a glance. Our volunteer veterinarian, her back pressed against the wall with a wall of men between us, looked like a deer caught in headlights. A lone Correction Officer stood guard in a fenced-off alcove. Where was our escort?

Tammy grabbed my arm. "I think they went down that hallway."

"Come on, Doc," I motioned. "This way."

That was the only time I felt a tinge of fear being inside. The inmates we worked with were like nephews, or grandsons. Tattooed men, guilty of crimes we tried not to think about, turned into eight-year-old boys melting with puppy licks when we presented their charges. Smiles grew.

Over time we noticed scraggly hair trimmed and shirts tucked in.

Tammy took on Baraga's new teams. I stepped in as puppy counselor at the Chippewa prison. Once a month, we traveled to meet with our respective inmate puppy raisers. Our trip was three days of driving almost 1100 miles, tending to the transport of seven-week-old puppies in or year-old-puppies out, and half-day training sessions.

Although the driving was tedious, the wildness of the UP was alluring. My friendship with Tammy grew and I got a crash course in developing training plans—with a first-hand view of the power of puppies.

"You and Tammy are role models for us," one inmate said. Hard to wrap my head around the notion.

"Don't underestimate that," another nodded.

The success of prison puppies created growing pains within LDB. As with any non-profit, staff was short, with significant reliance on volunteers. Lack of developed procedures and clear expectations from the organization created frustrations. Despite a passion for LDB's mission of "empowering people who are blind or visually impaired with lifelong skills for safe and independent daily travel;" despite a love for puppies (Tammy raised thirteen puppies for LDB; I've raised seven); despite a belief that the initiative affected the world in unexpected measures,

I became disillusioned. Inadequate support and communication from LDB led to feelings of being unheard and undervalued.

That my angst and aggravation with LDB (and yes, pride, too, for my teams) stayed home, is telling. That absence was not much different from when I took my first bicycle tour—I was gone an entire month with never one thought of the boyfriend I left behind. When I got home, it horrified me to hear he hopped on his own bicycle and rode it over 200 miles in one day. His legs cramped, and he called his dad to pick him up in Gaylord. Why did he do it? He never rode anywhere. If he was trying to impress me, he did not. I never missed him. We were done.

Until today, I didn't realize the weight of my volunteer engagement. Was I now done with the prison program? This trip, am I to learn something about how I'm living my life? Might pedaling alone be like searching for truth in a mirror, an unearthing of unconscious wants and needs?

Setting these thoughts aside, I yearn to be sun-like. Let be what be.

Thirty miles more of the Central Lakes State Trail to enjoy until the route returns to roads in Fergus Falls. At Evansville, I divert a block off trail to restock my food supply, and "second breakfast," an endearing practice of J. R. R. Tolkein's hobbits. The diminutive folk love eating seven meals a day: breakfast, second breakfast, elevenses, luncheon, afternoon tea, dinner, and supper.

Evansville, population 612, is a threadbare town. Nelson's Store, a throwback to an old-time general store, fills the corner of Main Street and Meeker. Displays of sewing goods, knick-knacks, and sundries with faded hangtags fill the front of the store. The smell of leather wafts from shelves of work shoes and cowboy boots. One side holds a meager produce section of tomatoes, onions, and potatoes; boxes and repurposed coolers heaped with goods not needing refrigeration crowd narrow aisles; the middle grocery section is only slightly better than a gas station convenience store (with a sizable choice of Hormel's Compleats, my new favorite ready-to-heat dinners). Tonight will be turkey and dressing with a two-serving package of garlic mashed potatoes. Also ending up in my hand basket are two bananas, a nectarine, and a yogurt.

The cashier, like so many other clerks I've met in this state, is cheery. She comments on the Hormel dinner. "Aren't these great? I think they are a take-off of those MREs the Army uses. They sure are tasty!"

I know she is referring to Meals Ready to Eat and nod my agreement.

We get to chatting and she finds out about my ride. "Michigan to Montana? By yourself? You are brave!"

"Maybe just a little crazy."

"Of course, around here you don't have to worry none."

I don't worry. Stuff can happen to you a block from home. I suppose the unknown causes fear, and people take a certain comfort in the familiar. Much the same as some of Mom's comments to me, I wish I could have a nickel for every time someone asks, "Are you going to carry a gun?"

I know nothing about gun safety and the rules for concealed carry and I cannot envision packing. Wouldn't a gun be heavy? And where would I keep it? A gun wouldn't do much good buried in a pannier if that Wisconsin bear decided to charge instead of running away.

It saddens me that guns seem to be a go-to solution. Are people prepared to suffer the consequence of taking another life? Not me. As a teen I delayed getting my driver's license because I couldn't bear the thought of killing someone in an accident. Cars aren't meant to be weapons, but that's how my young mind perceived them to be.

From now on, when strangers ask if I'm afraid, or say they are afraid to travel alone, I will reply, "Hey, everywhere I go, people say I don't have to worry about it here. So, I guess I'm safe everywhere."

A weathered bench in front of Nelson's is a perfect spot for my yogurt second breakfast. An elderly man pulls up in a '52 Studebaker.

"Hey Dale." The store owner steps off a ladder after straightening a bunting over the store window. They chat about classic cars and an upcoming parade.

"Nice ride," I say to Dale as he steps up the curb.

The deluge unleashes.

"I worked all over the Dakotas as an electrician. I helped build Indian reservations and missile sites. My wife was an angel, I lost her a year and a half ago. We had seven children and most of them live here 'bouts."

“My folks had seven kids too. Sorry about your wife, I lost my dad last summer.”

Dale looks off and brushes his pant leg. “Well, I’ve had a good life.”

Dad used to say that.

Evansville’s wide, empty streets remind me of another town. An image of Susan Notorangelo’s mother pops into my head. Years ago, I rode a RAAM qualifier in Illinois. The event, organized by Susan and her husband, Lon Haldeman (icons of the ultramarathon bicycling scene), was a figure eight course through dairy farmland. The route crossed in the town of Capron. Local women prepared hot meals under a big tent, a pleasant respite over the multiple-day ride.

At a picnic table, I enjoyed conversation with the women as much as their food. Susan’s mother, part of the food contingent and a staunch supporter of her daughter, approached. I had met her a few years earlier.

“You know Patti, you’d be pretty good if you stayed on your bike.”

I looked up at her admonishment thinking, *but I love talking to these people!*

I leave Nelson’s behind me, but the memory follows me back to the trail with a sense of déjà vu.

Like a curtain drawn, shade from trees lining the route gives way to a spotlight sun on the waters of Lake Christina and Pelican Lake and sparkles me blind. What are those birds posing like herons in the shallows? Bigger than seagulls, some are white, others are gray and brown, still others are black. I rub my eyes. What are those two swan-looking things with swords for bills? Pelicans? Yes. Pelicans. Their graceful flight over the path belies an awkward takeoff.

I don’t think I’m in Michigan anymore, Toto.

Shade closes back in, my eyes need a moment to readjust. A shadow pops from the brush. A fox. It trots ahead for at least a quarter mile, oblivious to the contraption niggling behind. A sudden sniff, I blink, and it disappears. In its place, the shape of a bicyclist appears.

A gray-bearded man on a Trek 520 touring bike rolls to a stop. The bike is traditional, complete with seasoned Ortlieb panniers, fenders, and a well-broken-in Brooks leather saddle. He looks tired and hot, a neon yellow jersey unzipped to mid-chest, but styling with matching socks and padded, fingerless riding gloves.

Vernon left Bismarck, North Dakota last week and is heading to the east coast. "I rode the TransAm in '88." The TransAmerica Trail is the original B'76 route. I share my story in return.

He leans in, peering at my recumbent. Nothing stock on this Tour Easy. I ordered the frameset (including seat and handlebars) from Easy Racers when we still had the bike store and assembled the rest myself. Any touring cyclist worth their salt will appreciate my custom specs: Shimano Ultegra crank with shorter (165mm) arms for my short legs; wide-ranging gears; Shimano nine-speed bar-end shifters; linear-pull brakes; and bomb-proof wheels I hand-built using Phil Wood sealed-bearing hubs, stainless straight-gauge spokes (forty rear, thirty-six front), and double-walled Mavic A719 rims topped with Schwalbe Marathon tires (700 x 37 rear, 20 x 1.5 front).

"Nice rig. Any mechanical problems?"

"Nothing so far." From the slump of his shoulders as he stands over his bike, I'm guessing he's wondering about the comfort of a recumbent. I'm relaxed, still seated.

"What kinds of speeds are you averaging? I've been doing nine to ten miles per hour."

I check my computer. "Today it looks like I'm doing just over twelve." He sighs. I add, "Must be the slight tailwind I've had the last twenty miles."

Poor fella, nothing seems to go his way. "I just can't eat good. I'm from South Carolina and we eat veggies!" He warns me about Fergus Falls. "Had a hard time following the ACA map." He perks up, "Be sure to stop at the Honey Hub in Gackle, North Dakota."

The Honey Hub is another cyclists-only camp and I do have plans to stop there. I tell him about Donn's Bicycle Bunkhouse in Dalbo. Vernon seems reluctant to move on; eventually he groans onto his Brooks saddle and we depart. I am grateful for the comfort of my recumbent.

First thing off the Central Lakes Trail at Fergus Falls is a Subway. I again take advantage of Tammy's gift card. Ah, so nice to sit in air conditioning for a while and refill my bottles with icy water. Next door is a Salvation Army resale shop. I still need another pair of baggy shorts that aren't as baggy as those I started with. Not much in clothing, so no luck.

Unlike Vernon, I have no trouble snaking my way through Fergus Falls, a larger town with a population over 13,000. What is challenging is the almost thirty sweltering miles of hills afterwards. Two days on graded rail trails has me spoiled.

Finally, Pelican Rapids, a quaint town of almost 2,500 residents. Sherin Memorial Park is a few blocks east of the main drag next to the city pool. An older couple, sitting on lawn chairs in front of their camper, wave hello. A closed-up tent sits forlorn at the edge of the Pelican River that flows around the park. No one else is around.

I call the phone number on the ACA map. "The camping fee is $10 for non-electric sites," a City Hall clerk says. "Leave it in the metal box on the sign at the entrance. Camp anywhere you like."

The perfect place is a flat spot of grass next to a pavilion and near a suspension bridge over the river. Breeze from the water feels like a caress. I park my butt on the top of a picnic table to FaceTime Mom. Tires crunching on the gravel drive catch my attention.

"What are you looking at?" Mom asks.

"It's another rider."

I turn the camera on my phone around so she can see the three-wheeled, low-riding Catrike. Must be the rider one day ahead of me from Donn's Bunkhouse. Either I'm making good mileage or he's taking it easy.

"Say hi to my mom," I tell the man. He waves.

"I'll let you go." Mom begs off. We blow each other kisses goodbye.

"Love you!"

"Love you more!" *She always said that to Dad.*

Eric is from southern California and is the rider from Donn's. "I'm riding from New York to Portland." We share ride statistics. He averages about fifty miles per day, with lots of rest days. "I stayed an extra day at the Bunkhouse. I laid around all day and read. I did this route a few years ago, but not on this bike."

"How do you like it?" I ask.

"I love this bike. Helps me climb. Doesn't matter how slow I go, it is easy to keep a straight line, and stopping to rest is just a matter of putting the brakes on."

I understand the advantage of a trike, smaller wheels mean lower gears, but I don't like its wide stance. Some of these country roads are narrow.

Eric shrugs off the camp fee and sets up his one-person, body bag of a tent on the opposite side of the pavilion. Can't really fault him on not paying the fee after skating for free my first night at Hog's Island. He has all the latest in gear and equipment but admits he doesn't cook. "I don't like healthy food." Quite a contrast from Vernon's stance earlier today. Eric asks, "How're the showers here?"

"Those other campers told me they're cold. I'm going to the pool." I'm hoping a gentle swim will rejuvenate my weary legs. And body. I've been watching my resting heart rate gradually rise from about sixty to sixty-nine.

The water and splashing rainbow of kids is refreshing. The area is rich in turkey farms. A huge processing plant employs much of the population, many of whom are Hispanic, African American, and Asian.

Diversity is one thing I miss after moving north from the Detroit-metropolitan area in 2011. Rural living is quiet and lovely, and white. The ACA route is mostly rural, and people greet me with nothing but openness and kind words. But given the divisiveness of today's political climate, what if I were a woman of color? How would I be received? Wasn't racism supposed to dissipate after the civil rights movement of the 1960s? It was no big deal having African American customers in our bike store, or as neighbors when we lived in the city. Yet, little more than a decade ago, an African American friend in a college poetry class opened my privileged eyes: "I can't even go into a store at the mall without a clerk shadowing me," she said. Racism rears its head daily in the lives of non-whites in America, regardless if I choose to see it or not.

The faded paint on the brick walls of the shower building is another déjà vu, flipping me to 1976. I didn't realize how many small towns in the west had city pools. Our group often camped in parks next to them. In Michigan you are never more than six miles away from a body of water, no pools needed.

Anyway, the pool shower? Hot. <smile>

A van with two dirt bikes secured on the back pulls in next to the tent by the river. A man and woman wave and stroll over. The couple is from Winnipeg, Canada. For adventure, they take their motorcycles to explore remote dirt roads. Curious about what Eric and I are doing and what kind of bicycles we have, the usual questions ensue. I don't tire answering. Reciting the same phrases without thinking is easy.

"Your tent is huge," the man exclaims. "It must be really heavy." I suppose next to Eric's little tube my tent might seem excessive for a single rider.

"You know, this tent is two or three pounds lighter than the tent I used in '76. People thought that tent was heavy, too, but I never got wet in it!"

After chatting a while, the man says, "You are the calmest lady I've ever met."

His comment surprises me. I am inclined to tell him this happens when you finish a hot, sixty-five-mile day on hills after riding almost two weeks straight. Instead, I thank him, my arms wrapped across my body, my mind fairly blank.

The woman says to her compatriot, "She reminds me of your sister."

"She does! My sister, Patti, is an ocean kayak instructor. She takes people on tours around Alberta Island."

"My name is patti, too." *Although I prefer not to capitalize it.*

"Wait. What? We call her Patti-Lou, though."

What are the chances? "I go by patti. My middle name is Louise." My mother's name.

The man steps back, twirls around, and slaps his thigh. "That's her middle name too!" He looks at his friend in disbelief. "This is too weird." He can't stop shaking his head.

"I'm not surprised. Stuff like this happens to me all the time."

"Can I get your picture, I want to send it to Patti-Lou right now." He pulls out his phone.

Of course.

STATS

64.58 miles, max speed 31, ride time 5:30, avg speed 11.7, TOTAL 812.72

Future Leader Dog Bear with an inmate raiser.

14

Deadline Day

Wednesday, June 22, 2016
Pelican Rapids, Minnesota to Fargo, North Dakota

One with the tree, the girl's bare knees knocked against deep wrinkled bark. Each foot, each hand found purchase. Leaves caressed her and she closed her eyes. When she opened them, she imagined it was like what she read in books, about being in the crow's nest of a rocking pirate ship, spotting for land.

She glanced down. Across the street, her mother sat on the top step of the family's cement porch.

"Mom!" The girl-warrior hugged the tree trunk with one arm and waved wild with the other.

The mother looked up. The sight of her daughter swaying at treetop didn't register at first.

"Mom! I can see Northland from here!"

Another gray hair erupted from the mother's scalp. Northland Mall was at least two miles away. "Patti! Get down from there!"

The sight of two picnic tables buried under all my gear makes a fun photo and post on Facebook: *Breaking news...a touring bicycle exploded in the city park this morning at Pelican Rapids, MN. Fortunately, everything was back in place within the hour.*

Eric's breakfast is a sweet roll. I offer him a cup of instant coffee and a pancake.

"No thanks."

North Dakota for both of us today. We agree to share a campsite at Lindenwood Park, a city park in Fargo.

"You'll get there before me," Eric says. "Just give them my name at the registration so I can find you."

Crisp sun and rain-cleansed air build confidence. Hills I would have had to walk up yesterday are no problem today. Cornfields rise into their own horizon above me, around me, touching the sky. Like something out of Star Wars, a helicopter lifts above the tassels and zooms low over the fields. Back and forth it frets, I've never seen a crop-dusting helicopter before. But then, I've never seen fields so vast before either.

Forty miles zing by. *Should I ride past Fargo?* The ACA route jogs north into the city, only to return south. Twelve miles less if I keep straight west. Just need to find a laundromat, I'm out of clean clothes. One nice thing about carrying a smart phone (when there is a signal) is the ability to search for things like laundromats. Ah, one ahead. I can skirt Fargo.

Eric. Ah, well, we didn't sign a contract, or even shake hands.

Crossing into North Dakota is a non-event, the road widens to a busy, suburban five-lane highway. I take sidewalk refuge to the laundry. Drat, dry-clean only. The clerk points me to a coin-operated laundromat downtown, near Linwood Park.

Broad clothes-folding tables are a good spot to spread out ACA and state maps to plan the next week or so. The route continues with section four of the Northern Tier, dipping south from Fargo to parallel Interstate 94 almost straight west to Montana. Yep, the same I-94 that crosses lower Michigan, a connection a few miles from my mother's apartment.

An ACA warning: LIMITED SERVICES FOR SEVENTY-EIGHT MILES BETWEEN GACKLE AND ENDERLIN.

Nothing like knocking me off my confidence horse. Fargo to Enderlin is about sixty miles. Lindenwood Park it is then tonight. Tomorrow I'll push over seventy miles to Little Yellowstone County Park, shortening the "limited services" distance to the Honey Hub in Gackle by eighteen miles. After Gackle I'll see how I feel and how the weather holds out. Maybe Bismarck by Sunday, a shorter day, less than forty miles, that day or the next.

Clean clothes and a plan ease my mind. A quick stop at a super K-Mart garners two days' supply of food, but no luck finding another pair of baggy shorts. At least my butt cheeks are healing.

Lindenwood is a shady spot along the Red River in the heart of Fargo.

Eric paid for our site and set his tube tent up under a small tree on a narrow RV campsite, between two fifth-wheel rigs that look like they've been stationed here for more than a few seasons. No sign of him or his Catrike.

First, FaceTime Mom.

"Patti! What happened in that park? All that stuff, blown up all over! Was that your bike?"

"What are you talking about?"

"The explosion. Are you okay?"

Ah. Ha! My morning Facebook post. "No, Mom, that was a joke."

"But all that stuff everywhere. What happened?"

"It was my gear spread out on two picnic tables. It looked funny, and I tried to make a joke."

"But…forget it." She cuts me short. Ah, well.

Second, call Andy. I leave a message, "I'm in North Dakota. Everything is fine. Call you later tonight."

Third, tent stretching. Ground tarp positioned with a slight elevation at head-end and unroll tent. Hamstring stretches to stake the corners, calf and lower back stretches to attach poles, lat stretch to shake out fly and toss it over. Quad stretches to unroll and blow up my Therma Rest pad, another set to unstuff and fluff my sleeping bag.

Fourth, dinner. While it heats, I document daily stats from the cyclo-computer. While I eat, I review tomorrow's route and add to my filling-up notebook.

Eric zips up on his unladen Catrike. He went for a shower. "It surprised me you weren't here yet." He climbs out of his low rider. "I figured you must've been flying with that tailwind and kept going."

"I considered it, but I rode all over town to find a Laundromat. By then I figured it was too late to keep going. Awesome ride today, for sure!"

"It's my birthday, I'm going into town for dinner."

"Happy birthday! Let me know if you find a good place for breakfast."

I'm glad for the alone time. This is my fourteenth day on the road, the end of my "at least two-weeks" demarcation. Andy waits to hear my decision. Have I had enough? Am I done yet? Should he come and get me? I will talk to him tonight, but for now I write my feelings on paper.

I ramble.

This ride. Not the simple "ride because I feel like it" ride I dreamed about for so long. No, it's something more.

For sure I am grieving Dad.

Years ago, my friend Linda told me, "I never knew a person so in touch with her own mortality as you." She crewed for Lou and me at a few ultra-cycling events. Did Linda refer to my propensity to "live in the moment" and not dwell on regrets or what-ifs? That grade-school student patti, who decorated the bulletin board with "This too shall pass" when the nun gave her permission to decorate how she wanted. The young woman patti, who found a job that paid enough for her "habits," in a field she could tolerate, instead of wasting time searching for a passion. "I know what I don't want to do," patti would say often enough. "I don't want to dress up every day or have a job that requires heels and make up. College seems a waste when you don't want to be anything." She just wanted to ride her bike. Working in a factory, when factory jobs paid union wages, was the right career. It paid enough to support her and the crazy ultra-marathon stuff she took to instead of riding off into the sunset.

I've been stuck on Gump "running until he was done." Was it the act of running itself that he chased? Or an exorcism of grief after his mother's death, or losing his girlfriend?

Waiting for a bus on a park bench, Gump told a stranger about his run, how he thought about his mother while running. And his friend, Bubba, who died in his arms in Vietnam, and Lieutenant Dan. (Gump carried Lieutenant Dan to safety after his legs were blown off.) "But most of all I thought about Jenny," Gump said. Jenny was the love of his life.

Perhaps exercising grief away is what I need, never mind what I thought I wanted.

Dad is in my mind every day. Him, reclined at night in his chair, me sleeping at the ready on the couch beside him. How he edged forward for a better glimpse of Mom in her nighty sneaking into the bathroom. Him, who said, "Say hi to Andy!" every time I left for a couple day's reprieve. Him, who sobbed, "They shoot horses, don't they, when they get like this?"

Hawks find me, daily it seems, in the almost eleven months since we've lost him.

Sometimes the beauty of the day hurts so bad I break into tears.

I am fortunate for this opportunity, for Andy and the space he gives me to leave for a while. I appreciate an older body that withstands what I ask it to do. I write to Andy how I've not been an overly introspective person throughout my life. I've been that person Linda knew, who lived each moment instead of analyzing everything.

Curious, now I am doing just that—my life in examination. Like Dad was in hospice, gazing beyond us, perhaps reflecting on his life, as the nurse said people do at the end. Every day a stretch of road throws me back to my past. Is this trip a vehicle for contemplating my life up to now, to reconcile the nearing end? Am I really in touch with my mortality as Linda said? Is that what's driven me to live my life as art—doing what I want in spite of social conventions, being who I am regardless, following a warrior drummer?

No regrets. Make mistakes but learn from them. What did Dad say when he took me for a spanking? "Don't be sorry, just don't do it again."

All my life, decisions caused me to ask myself, "What's the worst that can happen?" I'll die? We are all going to die anyway.

Perhaps Linda was right.

I sense I need something from this trip. I am *not* done. But perhaps the doing differs from what I imagined—riding as a way of life, to discover what is down an unfamiliar road, over the next hill, around the next curve. Have the years of competing in ultra-marathons spoiled me for laid-back ride-abouts?

I am pressed to press on.

Dear Andy, I must see this through. I will consider taking the train home, because maybe by the time I get to Missoula I will know what this ride is about, and maybe I will be done. Give me time to see.

Eric returns as I tuck Andy's letter into its envelope. I'm eating sliced peaches for dessert. "Want some?"

"No, thanks. I got a free ice cream for my birthday. Randy's Restaurant, right on University Ave a few blocks away. I'm sure it'll be a good breakfast place."

Dusk at Lindenwood brings a parade of work pickup trucks, pairs and more of young men ending a long workday in the oil fields. The worn RVs and campers stationed around the park must be housing for itinerant workers.

Two white trucks, with welding equipment in the beds, pull into the site next to us. Five grimy men get out and gather at the picnic table. One of them retrieves a cooler from the fifth-wheel camper, and hands cold beers to the others. I wonder if drunken racket will spoil the night. I am glad Eric camped with me. For the first time, I feel a bit of unease traveling alone.

My fears are unfounded. The men are quiet, as dark settles they retire.

Before sleep, I call Andy.

"It's your fourteenth day," he broaches.

"I wrote you a letter. You may hope that I am done, but I am not. I must get to Missoula at least."

"I'm not surprised."

Our talk rolls to the mundane. He tells me how my friend, Mim, who lives a couple miles down the road, brought him fresh chicken eggs and homemade tortilla soup. "She put my day of reckoning off one more day." No leftovers remain in the freezer. He didn't want me to leave enough for the entire summer. "I can take care of myself." But others are helping. Our friends, Phyllis and Dick, have him for dinner once a week.

I fall asleep relieved and ready to begin the "real" journey. No matter how much you prepare, the first week trains the body. The second week shifts the mind to a lifestyle on the road.

POSTCARD FROM THE ROAD 6/22/16

The moment I clip my right shoe into my pedal and push off, my left foot leaves the ground and I am five years old again, feeling freedom for the first time.

Tires sing; the long chain hums, lifting into the big chainring like a hungry tiger. Spinning is effortless, the air crisp and light, and I am flying.

STATS
63.97 miles, max speed 35, ride time 5:13, avg speed 12.2, TOTAL 876.11 Average mileage for 14 days: 62. I'm on track.

15

Limited Services

Thursday, June 23, 2016
Fargo to Little Yellowstone Park, North Dakota

Saturday evening, the girl was overwhelmed. An experienced babysitter at eleven years of age, this gig (two doors from home) started Friday night after dinner. There were two kids under seven; she'd taken care of them before. They behaved well enough, but she didn't know how she would make it until the parents returned Sunday afternoon.

The phone rang. Her mother. "How are you doing, Patti?"

The girl broke down. "I can't do this!"

"I'll be right there."

After settling the kids in bed, the girl and her mother raided the potato chips and Coke (left just for her) and curled up on the couch together. Time for the Andy Williams Show, the girl's favorite.

The workers are up and gone before I crawl out of my sleeping bag. I don't envy them their long day. Won't see Eric again, he's taking a short day into Kindred, less than thirty miles away. "I wanted to go to Enderlin," he said last night, "but the town is having a reunion or something. I called a hotel. No rooms."

I shake the morning cool at Randy's Diner with my usual. I can't eat it all. Never a problem eating before this trip; weight-gain followed a ten-year high and low tide. A dripping 100 pounds at the start of B'76, my five-foot-two frame was twelve pounds heavier at the end. Rationalized as muscle gain, more likely it was the all-you-can-eat buffets, ice cream, and country pie contests along the way.

Racing weight as I approached thirty hovered a bit below 120. At a cousin's wedding reception, the afternoon after a sprint triathlon, a woman twice my weight lifted her nose. "Why in the world would you want to do something like that?"

"So I can eat anything I want," I snarked, diving into a Vesuvius of finger food balanced on a teacup sized plate. Who serves hors d'oeuvres at a wedding?

By fifty? One hundred fifty pounds. That year, Lou, evil-tandem-twin she is, whipped me back into shape (and down to 120) to ride 147 miles from Lake Michigan to Lake Huron in the ODRAM (One Day Ride Across Michigan). Twelve brutal hours against head winds. What is this myth about west winds? Ah yes, it's me, the Headwind Queen..

Last December, peaked at 163, I scheduled the physical I put off for two years. I may be crazy, but I calculate risks. I needed to be sure I wouldn't drop dead of a heart attack.

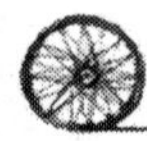

The physician assistant was not happy with my cholesterol and blood pressure, never mind my weight. "You need to be on statins." Given my family history of heart problems I suppose she had a right to lecture. Both parents had by-pass surgery and two brothers survived heart attacks.

"I won't take them. I know what I need to do, and I'll do it. But there's something else. I want to ride my bicycle to Montana this summer."

"That's it. You're going to have a stress test. If you're lucky, you'll test positive and have to have a catheterization. That way you'll know the condition of your heart for sure."

The technician ran a preliminary EKG before the treadmill test. "I'll be right back." She left me cold on the exam table for twenty minutes. "I talked to the cardiologist. We can't do this today. You'll have to come back for a nuclear test."

"Why not?"

"The only thing I can tell you is if we did the test today, it would be un-diagnostic."

"What does that mean?"

"I can't tell you anything more."

Why? If you told me would you have to kill me?

A few days before my rescheduled test, the nuclear picture-taker called. "Fast the night before. And wear a bra."

"A bra? I haven't worn a bra since I was twelve!"

He stammered and mumbled something about breast cancer.

"Nope, never had cancer. I just don't wear them." Much to my mother's chagrin. "You can see your nipples," she'd say. "Everyone has them," I'd counter.

Picture-taker insisted he couldn't do the test unless I wore one. A tank top wouldn't do. He finally agreed to a jog-bra. "And wear your husband's flannel shirt. My room is cold."

I have my own flannel shirt, mind you.

What was I in for? An ordeal, it turns out. From a lead-lined syringe, a nurse injected radioactive fluid into an IV port in my arm. So as not to expose anyone to the radioactivity coursing through my veins, I sat alone in a special room. For forty-five minutes. Finally, flat on my back on a table, an MRI-like machine encircled me. "Hold your breath every time you hear a bang," the technician instructed. That meant every fifteen seconds when the machine changed position.

"Do you want chips and a Coke or a pastry and coffee?" the technician asked as I slid off the table.

"You're testing my heart and you want to give me junk like that?"

"The fat helps us get better pictures and the caffeine will help you avoid a headache."

"I'll take the coffee and pastry."

More time in the waiting room, injected with medication to speed up my heart, and more time on my back for another round of pictures. A second technician, sitting at a computer screen remarked, "I wish all of these were this good."

"Are you referring to the quality of the pictures or my heart?"

He shot a glance at the other tech. "You can go now."

That was something at least. Mom was sixty when she had a nuclear stress test. A 90% blockage sent her straight to the hospital for surgery.

Two long weeks later a postcard came in the mail. "Normal." I guess my heart could take the stress of a bike ride across the country. But why a bra? I never even unbuttoned my flannel shirt.

I call Andy during breakfast. He says, "I've had a terrible morning. I dropped the glass soap dispenser Jen gave me and I had to clean the entire kitchen floor."

I picture him on his hands and knees, swearing at the unexpected task. "Ah, well, got a clean floor out of it."

He gives me the fence report: he finished the pickets between the house and his shop, today he will make the gate for that section. Before I left, I helped dig sixty-four holes for cedar posts, by the time I get home our backyard will allow Future Leader Dog puppies to run off-leash.

"I've been looking at your mileage. If you keep on pace, you'll have 300 miles to spare by July 15."

I don't share his confidence. Aren't the plains notorious for winds? And I expect I'll be doing a fair share of pushing my load up the mountains. If I have to, I can rent a car to make it to the celebration dinner, but I'm determined to ride.

Self-imposed pressure, my M.O.

Maybe Andy brought me good luck. Tailwinds out of Fargo lift me to Kindred, "Where kindness is a way of life." I take a break at a convenience store in a gas station and buy a vanilla yogurt parfait with granola and berries. The picnic table on a small patch of grass outside is inviting. At first. The sun beats down on nearby pavement, ricochets off the white-washed walls of the store and boils my brain. Like a puppy searching for water, I head to the air-conditioning. I hadn't noticed the group of tables at one end of the store and melt into the nearest seat.

Tomorrow is Dad's birthday. I text my younger sister Anne to see if she wants to go in on flowers to send to Mom. I'm sure it will be a hard day.

A local sheriff struts over, followed by a young man dressed in jeans and a plaid shirt. "We're going to sit behind you and protect you," the sheriff says. I thank him when I head back to the inferno.

A few miles down the road he passes me. I wave and he flashes his lights. I think of my youngest brother, Joe, a Dearborn, Michigan cop, putting his life on the line every day. He wanted Dad to see him retire, a few years away now. I hope Joe makes it. We almost lost him to a major heart attack when he was forty-five.

"Hang in there, brother," I whisper, watching the sheriff car disappear in the distance.

The road heads straight west now, the wind no longer my friend. Ahead, a cyclist shimmers into form. I am disinclined to interrupt my pace so soon after a break, but he gestures a stop.

We are instant comrades.

Dave is a lanky man sporting a white beard. His bicycle is so tall it makes his front and rear panniers look out of proportion. Retired, a month ago he left his home in Seattle to ride to Maine. He feels his own deadline pressure. "I have to get there by Labor Day. My son is getting married and I have to fly back."

We share road gossip.

"I stayed in Enderlin last night," he says. "The town is celebrating its 125th anniversary, and they put on an amazing performance of the 'Music Man.'"

"So that's what was going on. The rider behind me said he couldn't find a room."

"Two women are behind me, from Oregon heading to the east coast. I've been riding off and on with them. You've got some road construction ahead. It's passable, but not fun."

Great.

Further on a sign confirms Dave's warning. "Road Construction 13 miles." No evidence for miles, except for three short sections of gravel interrupting the pavement. Eventually, a woman wearing a hardhat and bright orange and yellow garb stands in the road with a stop sign, holding traffic. She calls for me to stop and gets on her radio. "I have a cyclist here, what should I do?"

Just then, two women (I assume the ones Dave had told me about) cruise past. Their shout to me startles the worker. "You're the first woman we've seen!" Without stopping, they yell over their shoulders, "We've got mail in Kindred and the post office closes at two." It is only noon. They have plenty of time to make it, but I understand their reluctance to pause.

The woman says, "It's your choice. You can try riding the two miles, just watch out for the trucks. Or you can take a ride in my driver's truck."

It's hot, the road is a mess. The lead truck arrives followed by a long line of traffic. "I'll take the ride, but it will be hard to get my bike in the back of his truck. It's heavy."

"We'll get it in." Sure enough, between her and the driver we muscle it.

"I can ride back here," I say to the driver.

"Nope, safety first. You ride up front." He shoves papers and clothes off the passenger seat. "I live in this thing twelve to sixteen hours a day. We're building a shoulder, taking an inch off and adding an inch and a half, so next time you ride through here it will be nice."

I wonder if I will ride back this way.

Phineas is talkative. "I normally put in 400-600 miles a day on construction sites like this. My best day was 650. I'm originally from the south but we've lived here in North Dakota for twenty-five years. My wife is still not used to the cold."

Getting the bike out is easier than getting it in. "Drink lots of water," Phineas advises as I step over to let traffic file by.

At Enderlin, the last outpost for seventy-eight miles to Gackle, I explore off route for supplies. A wooden sign ("The Organic Cupboard"), shaped like an antique cupboard, stands in front of a white bungalow which houses a massage and chiropractic business. Two side panels of the sign are open, chalkboards advertise local and organic fruits and vegetables. "Fresh Peaches" captures my eye.

The inside of the store presents like a Paul Cézanne still life with vegetables and fruit spilling from woven baskets. Organic pastas and gluten-free breads line shiny metal shelves. Refrigerated cases along the walls hold frozen grass-fed beef and pork and dairy-free ice cream.

A store like this in a fewer-than-900-residents town? I wander wide-eyed and set my selections on the front counter, behind which a brown-skinned boy about eight years old perches on a stool. Short tight curls ring his bright face.

He hops to his feet. "I'll get my mom to help you." He returns dragging a young white woman wearing an apron.

"Can I help you?"

"I can't believe there's an organic store here."

"Doctor Maggie Peterson, the chiropractor next door, started the store in 2008 for her patients and the community. I moved here from Georgia two years ago. I miss trees."

"Me too. I'm from Michigan."

The boy pipes up, "I've always wanted to visit Michigan."

"You should. It's a beautiful place. Lots of trees and water."

He follows me out and two other kids run over to examine my bike. The boy admires my fairing. "That is so cool."

I pedal away smiling like I won the lottery. Poor Vernon, the rider I met on the bike trail in Minnesota who complained about food choices, he missed an awesome store.

The last fifteen miles to Little Yellowstone Park on the Sheyenne River turns hilly. Of course. The ACA map describes riding conditions across North Dakota as "a vast, open plain that slopes downward from the Rocky Mountains to the Mississippi River." From the east, 900 feet of elevation at Fargo rises to more than 1800 feet at Bismarck, and 120 miles further west the town of Dickinson tops 2600 feet. A bit of climbing, sure.

Around a gentle curve, the road approaches a green line of trees (*the river?*) and drops as if descending a mountain. I work the brakes, not because I fear speed, but I don't want to backtrack uphill if I miss the park. Instant regret at the bottom—the entrance is at a natural breaking up sweep. Let tomorrow concern itself with the climb out, shade from stately trees provides an oasis from the open prairie tonight.

A self-register kiosk invites campers to set up at primitive or electric sites. A tent-site for me is $15, but all I have is a $20 bill. No other campers, I stop a couple walking around and ask if they have change.

"Sorry, no." The man tells me they came to check out the park they visited as kids. "We had so much fun here exploring along the river."

I pull an Eric. I'm willing to pay, but my budget is tight, and I don't want to waste five dollars.

With no cell service, I walk up the hill on the road for enough signal to text Rick. It's Thursday and he'll be having dinner with Mom. Each week they pick a different nationality restaurant. She'll worry if she doesn't hear from me.

Success. "No cell service. Tell Mom I'll talk to her tomorrow, maybe late." I get an email off to Andy so he doesn't worry either.

A family drives up in a minivan. Mom and dad take three young children out of car seats to play on the swings. The woman comes over to chat. "My husband played here as a kid. We wanted to bring our kids here to see the park."

The children follow their dad like giggling ducks, clambering down the root-mazed, black dirt bank of a stream that flows from the Sheyenne River.

I sense a theme.

The stream is murky, pitiful compared to the holy waters of Michigan, yet holds an allure for generations of its residents. What is it about moving water that is so compelling? We are nothing but a drop among uncountable others. Sometimes we travel far, other times we get caught in an eddy to circle until we don't know where we are or where we should go; sometimes we float in peace, other times we feel like we are drowning.

What droplet am I today? Surely I travel with purpose to get to Missoula, even as grief washes over me. Yet something more than a simple Forrest Gump ride looms ahead.

As if to match my dread, a strange bellowing overshadows the rumble of trucks barreling down to the river and up the other side. Barrooo! Primal and frightening, the creature sounds in deep distress, like a cow in full labor or a roped bull. Hair pricking at my neck, I wander across the stream and follow a tight trail to more tent sites.

BARROOO! Across the dirt road a barbed-wire fence holds brush and wild at bay along a steep rise. *What's that shaking, a tree or a bush? Is that a cow caught in something?* Light is too dim and I am not bold enough to manage a closer look.

There—black cows. BARROOO, BARROOO! Nope, the stuff-of-nightmare howls isn't them. I back away unsettled.

STATS
71.42 miles, max speed 26, ride time 5:15, avg speed 13.5, TOTAL 948.15 I might be halfway!

16

Remembering the Ride of my Life

Friday, June 24, 2016
Little Yellowstone Park to Gackle, North Dakota

"Tell me again how you almost lost me," the girl begged.

"You scared the heck out of me," her mother said. "Out of the blue one morning, your diaper was soaked bright red."

"How old was I?"

"Nine-months. The doctor wasn't sure what was wrong, maybe allergies."

"What did you do?"

"I said a novena."

The girl didn't understand novenas, something about intense praying. She imagined her mother on her knees, bowed over her rosary days at a time.

"And then, just like that, you were better."

Dark clouds loom with an annoying drizzle for forty hilly miles. They part and the sun heats steamy wind in my face. At the top of a long hill, I put my feet down. Listen. Wheat feathers rustle like a rain stick. Look. Waves of wheat under rippling cloud shadows are so lovely I take an iPhone video. As I pan, three riders rise from the west like Viking ships.

With grins as wide as the terrain we're riding through, they tell me they head all the way to Enderlin today. "It's supposed to be windy tomorrow."

What do you think it is doing today? Oh, right, you have a tailwind.

One rider is a young man from Maryland on his first tour from the west coast to home. Sporting a slick red, white, and blue Lycra jersey he straddles a carbon fiber Trek racing bike, his gear stashed in a cargo-trailer under a yellow rain cover. His Hawaiian friend (the only one without a helmet) tows a makeshift trailer that looks suspiciously like a golf cart. Ablaze in an orange safety vest, the third man chides the rookie over a run of flat tires.

"Look at hers," he says in an English accent. He points to my Schwalbe Marathons. "I've been telling you, you need touring tires."

The Trek's narrow, high-pressure tires, meant for speed, are not so good at repelling road debris.

"Good for you, taking a trip with what you have," I encourage.

With mirrors extending a forearm length from both sides of multi-position handlebars, the Englishman's scruffy hybrid bike resembles a loaded semi. Fenders, a can of bear spray strapped to the frame, and bulging front and rear panniers well used from more than one adventure—this man is not a rookie.

The two friends met the Englishman a few days ago, he's been on the road for a while. From the east coast, he rode west on the ACA's Southern Tier route, found his way up the coast to Washington, and is taking the ACA Northern Tier east.

The rookie grumbles about wind and hills.

"You've been losing altitude since Montana," I tease. "It's all downhill from here."

The English rider says, "Wind just means it takes twice as long to get as far. You work just as hard, you just go slower."

The Hawaiian man adds, "You need to make peace with the wind."

I laugh. "The good thing about hills is they break the wind for you while you're climbing."

The Hawaiian points at me. "I like her attitude!"

The three are amicable companions. I'm a bit sorry they aren't traveling in my direction. On our separate ways, I hope they enjoy the same energy spurt I get from our encounter. My eleven miles per hour increases to fourteen.

And yet. The forever-straight road, the beating headwind, the endless fields that roll like the ocean, swells my mind to sorrow. It's Dad's birthday today, he would have been eighty-nine.

Three miles from Gackle a dip in the road between two side-rising hills gives a boost against the wind. A bee zings a near miss, reminding me of the Matrix-bee in Minnesota. Two bees, one bounces off my fairing. Here a bee, there a bee. They're everywhere.

These bees do not buzz in slow motion, they scatter in desperate alarm, smacking against fairing and helmet. One lands on my arm. I brush it away. Where are they coming from? At the gully between the hills, hundreds of peeling white bee boxes are stacked in a jumble some twenty yards from the road. Swarming is heaviest here. As I ascend, the wind whips them away and I escape unscathed.

The ACA map noted to call Jason at the "Honey Hub of Gackle," a cyclist-only lodging. When I called yesterday, a recorded message gave directions to the house and an open invitation to make myself at home. Cruising through the small town, I pass a woman talking to an older couple in the driveway of a corner house.

The woman waves. "Are you heading to the Honey Hub?" I pull over to meet Ginny, Jason's wife, and her parents. "Keep on for two blocks, turn left, and it's the white house at the bottom of the hill."

Easy to find. A board painted white points to the back of the house where a short-wheel-base recumbent leans against a redwood picnic table. A clothesline stretches across the yard that slopes up and away, dotted with a few trees.

The Honey Hub is an outside entrance basement room with two beds, couch, hotplate, microwave, toilet and shower, and washer and dryer. A man about my age sprawls on a bed. Barry is riding from Washington to Raleigh, North Carolina. This is his second transcontinental, he rode a different route in 2008.

"I wrote a book about that ride." Barry hands me a business card. "And I'm keeping a blog bout this trip. Are you staying here?"

"Yes."

"Inside?"

"Nah, I'll set up outside." A shower and a chance to wash clothes and charge my phone is all the inside I want. Did Barry breathe a sigh of relief?

A knock from an inner door and Jason sticks his head in to introduce himself and welcome us. Born and raised in Gackle, he owns the house and his family's multi-generation beekeeping business. He winters in California with his bees and brings them to North Dakota for the summer.

"I race triathlons. When I found out about Adventure Cycling routes through Gackle, and my house empty much of the time, I wanted to do something for the riders."

I ask about the bees that swarmed me a few miles away. "Yep, those are my bees. One of our trucks unloaded 500 hives there today. They'll be gone tomorrow, but we had to use a spot close to the road for the semi to drop them."

"How many hives do you have?"

"About 25,000." Holy moly, that's a lot of semi-loads.

The grassy yard is comfortable, I sprawl on my Thermarest pad in the shade. Free WiFi. An email from Andy! He says he reads my Facebook posts and the comments that follow. He says, "You should try writing!"

Love this guy, who bought me a laptop our first Christmas as husband and wife. "For your book," he said. Will this trip be the story I finish? Everyone is a writer these days. I will write my book for me, and any friends and family who might be interested.

Barry surprises me with a dinner invitation, pizza at the only bar in town. "My treat," he says.

How can I turn down free food?

We walk the few blocks to the bar. Closed. Fortunately (or not), the Gackle Tasty Freeze is down the street. Jason warned us about the proprietor: "She doesn't like strangers."

Jason is right. The woman wears a perpetual Grinch smirk. When we get the frozen pizza that Barry ordered and the salad I bought to share, we ask for an extra plate and silverware. She slides the food across the table, stomps off, and returns to slam plates and silverware. Our "thank you" falls on deaf ears. Ah, well.

"It's my dad's birthday today." I tell Barry about our ride at a twenty-hour race in 1986. I don't care if he gives a hoot, I feel a need to remember.

At the end of our failed attempt to complete the John Marino Open on the tandem, Susan Notorangelo, organizer of the event, tried to console my dejection. "You should do the Michigan National 24 Hour Challenge. The women's record isn't very high."

"I know. I set it." In 1984, I rode 324 miles for the win. Susan had nothing left to do but pull her foot out of her mouth.

The next year, and many training miles later, a more confident me planned bigger mileage. I talked Dad into coming. "Dad, if you can ride at least to the first checkpoint, only fifty miles, I'm sure we can win the father/daughter category."

The Michigan National is held on Father's Day weekend. Dad rode his bicycle to work for years. He and Mom sometimes rode thirty or forty miles in club events such as the Peach of a Ride or the Blue Water Ramble. I added a tall stem and wide upright handlebars to my Raleigh Super Course (the bike I rode in B'76) to give Dad a more upright riding position, and we did training rides on my tandem.

The three-route event started at 8:00 a.m. on Saturday morning. The first loop was 115 miles around Grand Rapids. Riders then had to complete at least one twenty-six-mile second loop before dark and the six-mile night loop.

I figured Dad might make one or two checkpoints, but as I came off my third twenty-six-mile loop, I caught Dad rolling into the main checkpoint. My smile beamed ear to ear, thrilled he completed his first century. His eyes torpedoed me with a snarl, "You lied."

Yes, I admit I told him it was not a hilly course. I grinned at him, "Lying is legal."

Pissed or not, Dad cheered me on from the sidelines. Susan, riding to prepare for her upcoming entry in RAAM, held a small lead into the night. Her husband, Lon, asked Dad, "What is Patti doing?"

"Riding her butt off!" His daughter was a threat to this athlete he saw on television.

Rain started during the day and kept on after dark. A car hit a tandem, no serious injuries, but they shut the course down for over an hour. When it reopened, Susan decided to not risk injury and dropped.

It was my race.

My high deflated with sunrise. Only full loops counted when the clock hit the twenty-four-hour mark. With only twenty minutes left, I was ready to stop.

"I'm done. No time for another loop."

"Get back on the bike," Dad said. "You can do it."

I couldn't argue. I kicked my pedal around to remount and told my riding friend, Mike Dobies, who had been pacing me, "Make sure I get back before eight."

Head down, I followed Mike's wheel. He got me there in time.

"It was the ride of my life," I say to Barry. "My face hurt from smiling. I ended up winning with 379 miles. And we took the father/daughter award."

We do hard things like pushing our bodies for twenty-four hours. Or across the country. We might feel like we are dying, and yet we don't.

Dying is real.

Later at the Honey Hub, Barry asks, "How often should I lube my chain? I haven't lubed it since I left." He left the west coast at the end of May. It takes only a glance to see that his chain needs lubrication. Bad.

"I lube my chain every time it gets wet, and whenever it gets a little noisy. Your chain looks like it needs it."

"How do you lube your chain?"

I explain. He retreats and returns with lube and a roll of toilet paper.

"Don't use that!"

"It's the only thing I could find."

"Wait, I've got a shop rag. I'll do it for you." I rotate his crank backwards to drip T-9 (my lube of choice) on each link. The rear derailleur pulleys chirp like a flock of crazed grackles. "Doesn't this noise drive you crazy?"

"I thought it was normal."

I proceed with a modified "Michigander tune-up."

It is nigh impossible for me to get Andy off an idea once he thinks it's a good one. The success of my providing technical support at the 1995 Michigander Bicycle Tour gave him the idea for me to promote our store at events around the state. But he didn't want me sleeping on the ground, not that it mattered to me. On a wintry March evening, we picked up a thirty-two-foot Bounder motorhome from a dealer in Belleville, about an hour's drive from our place in Clinton Township. The treacherous drive home on snow-covered roads was a good initiation.

We gutted the inside and converted it into the "Rover," a store-on-wheels. My artist friend, Linda, designed a mural to wrap it. For nine years, I drove the Rover on tours to work on bicycles and sold enough swag to cover its cost. In the spirit of promotion, I only charged for parts. Some took advantage, but I did what I could to keep riders riding for the day, weekend, or week. After long days, I buried my sore hands in the ice cooler for relief.

My friend, Debbie, was a godsend sidekick when she accompanied me on several events. She oversaw check-in, so people didn't have to stand in line for hours, although the "liar's bench" around my work stand was always full. Tall tales flew as I wrenched, and astute bike bums learned a thing or two.

Most bikes got what I dubbed "the Michigander tune-up," a quick check and adjustment of bearings (hub, headset, and crank), brakes, shifting, wheels, and tires; a wrench on every bolt to assure proper torque; and lube of cables, chain, and moving parts.

Barry bends over me. "I've been having some shifting problems, do you know what causes it?"

"It's likely the derailleur housing." I squirt lube into the pulleys to quiet the chirps.

"This thing?" He points to the cassette, the gears on the rear hub.

Inside my head I shake it. For a guy who rode across the country once, and is well into a second crossing, Barry has no working knowledge of his bike. A shortcoming, especially when riding a recumbent. Many bike shops are not familiar with the idiosyncrasies of these bikes and rarely carry extra-long cables or other parts peculiar to them. I'd bet the English fellow I met this afternoon knows his way around his bike.

"Nope, here." I touch the small loop of plastic housing the cable.

I backpedal to wipe off excess lube and ask Barry to hold up the rear end so I can pedal forward. This is my Michigander tune-up secret: shift to the inner rear gear, stop pedaling and release tension on the cable; the chain holds the derailleur so I can disconnect the loop of housing from the cable stop; slide the housing up the cable to expose the section with corrosion (causing delayed shifting) and lube it; check derailleur limit screws (they prevent it from moving too far into the wheel or too far the other way) before reattaching the housing. Finish with a few drops of lube on the pivot points.

"Your shifting should work better now."

To be honest, I didn't know much on my first tour in 1974, or the next summer with my brother. B'76 taught me a lot. The second morning our assistant leader gave a lecture on checking our bikes every day before heading out of camp. Mike worked in a California bike store and was our most experienced mechanic.

"First, check your tires," he said, giving his front and rear tires a squeeze. "Then lift each one and give it a spin to make sure the wheel is running round and not rubbing the brakes."

My tires were fine, front wheel fine. My back wheel wobbled and stopped against the brake pad. A broken spoke. Of course I didn't have a spare. Why would I? The only mechanical problem in my earlier tours was one flat tire. Mike trued my wheel enough for me to ride through Oregon and part of Idaho. Three more spokes broke before Lolo Pass into Montana. During a rest day in Missoula, the famous Sam Braxton of Braxton's Bike Shop rebuilt my rear wheel and installed a lower gear to help with climbing.

Mike kept Len and Loree's Gitane tandem running (and everyone else's bikes) until his sore knees forced him to leave the ride in Colorado. After that, Bob took over. He rebuilt the tandem's rear wheel on the sidewalk in front of a bike store somewhere in Kansas. Watch and learn time for me that summer.

Several years later, when Lou and I were new to the tandem, our front derailleur cable pulled free during a critical shift in a time trial. A bike shop serviced the bike before the race. I vowed to never again trust anyone to work on my bikes and obsessed over keeping them sound. Before every long ride I give them a once over, perhaps the original start of my "Michigander tune-up."

In 1987, Lou and I became the first women's tandem team to complete the Paris-Brest-Paris (PBP), the oldest "brevet" in the world. (Brevets are long-distance cycling events with designated checkpoints and a time limit.) Originating in 1891, the 1200-kilometer PBP (since 1975) is held every four years. To qualify for entry, Lou and I rode four progressively longer qualifying brevets in the United States that same year (200, 300, 400, and 600 kilometers).

In the first couple hundred kilometers of PBP, we experienced weird mechanical issues, like losing a bolt from our rear rack (I carried spares), or excessive cable stretch causing poor shifting. At first, I attributed it to rough pavement.

We always drew a crowd, the infamous "Léopards Roses" ("pink leopards") because of our gaudy, hot-pink, spotted tights. I started giving the tandem a quick check before leaving. At one stop I found a wobble in our rear wheel. Lesson learned in B'76, I had extra spokes. No spokes broken, but one was completely loose. This doesn't happen by itself.

A spectator gestured, "Mécanicien?"

Do I want to take my bike to a mechanic? "No!"

Lou lifted the rear end, and I set to work with a spoke wrench. The crowd parted to let us leave.

Lou said, "I think someone was testing us. This never happens. You take such good care of the bike, someone had to have messed with it."

Was she right? Were we being tested? If so, we passed. We had no problems the rest of the ride.

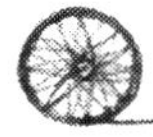

I crawl into my tent early to call Andy.

"Last night was dinner with the Downings," he says. "I took them to the Clear Lake Bar. This morning I worked with Seth cutting logs, and we'll be out there again on Monday with Trinity or Jory." The three are sons of my friend, Mim, who brought Andy eggs and soup. "John's coming to fish on Tuesday." John is Andy's friend since fourth grade. "And I've figured out a solution to watering the garden from the slow rain barrels. I bought a cheap pump from Harbor Freight and hooked it to the boat battery. It works great, only takes fifteen minutes to empty a barrel."

I hear his triumphant smile. "No fair," I say. "You could have done that for me years ago, how many mornings did you watch me water for over an hour with no-see-ums buzzing my head?"

"Sorry. Not sorry."

"Ah, well, it was meditative."

"It really hit me last night," he blurts. "Not being able to talk to you."

The subject of me being done comes up again. "If I take the train home after the celebration, I could be home in time to drive with you to the wedding." His niece's wedding is in early August in Maryland. "Remember our first trip to Kwagama Lake? You said you were excited to spend long hours trapped in the car with me." The cross-country ski resort, 118 1/2 miles north of Sault Ste. Marie, Ontario via the Algoma "snow train," was a favorite of mine; we took a long weekend there the first winter we dated. "Well, I admit I'm attracted to being trapped with you to Maryland."

"I leave that up to you."

POSTCARD FROM THE ROAD 6/24/16

When your ACA map warns of limited services for a seventy-eight-mile stretch between Enderlin and Gackle, ND you'd best be sure to have food and water to make it through.

A roll of TP is another necessity. The rolling, open fields don't allow for cover. While traffic is scarce, you can see matchbox trucks from more than two miles away.

Twenty miles from camp I felt the need. Finally, I saw an overgrown two-track run north from the road with a tree about 100 yards beyond. I pulled in and I clipped the straps on my pannier to retrieve the TP.

A four-wheeler whips out of nowhere from the west and almost lifts two wheels making the turn into the very same two-track. Two guys wave hello as they zoom past about twenty feet, jump out, and proceed to work on a fence overgrown by prairie grasses.

By the time I restrapped my pannier for an unrequited departure, they hopped back in and raced off. I have no idea where they went, they just seemed to disappear, but by now I was resigned to find another rest area.

Three miles later, a better spot...

STATS
58.63 miles, max speed 31.5, ride time 4:33, avg speed 12.8. TOTAL 1006.79

When I met the three guys riding on the plains, I thought I messed up my computer. It only read 6.79. Turns out the total mileage cannot register 1000.

Dad and me, 1986

The lineup for repair at our Rover. One of many Michigander bicycle tours in the late 1990s.

17

Surfing the Plains of North Dakota

Saturday, June 25, 2016
Gackle to Hazelton, North Dakota

A folded Kleenex held in place with a rubber band around her head dammed the Tahquamenon Falls of her allergy nose while reading. Her mind, at least, traveled on words in books she checked out from the Detroit Library branch on Seven Mile Road every Saturday.

She breathed best moving.

When she learned the earth travels around the sun at 67,000 miles per hour, in an ever-expanding universe, never to occupy the exact place in space ever again, the girl grinned. In motion whilst sitting.

My 5:00 a.m. bladder is now a 4:00 a.m. bladder. Damn time change. I hold steady, not wanting to wake Barry with a creak of the screen door into the Honey Hub. To take my mind off, I stretch and flex my body mechanical in morning assessment.

Toes point down, no cramping. Toes point up, check. Knees press flat, rock-on-rock quads. Pull knees to chest and roll side to side, lower back loose. One at a time legs reach for the sky. Almost. I did not inherit my mother's flexibility.

Feeling ribs brings to mind a thirty-some-years-ago massage by a Belgian man who raced with Eddy Merckx back in the day, when the drugs of choice for Tour de France racers were cigarettes and wine. The therapist dug his fingers across my intercostals. "No fat here." This morning the iliac crest of my pelvis rises like hillocks above the now-low prairie of my belly. Is this thin me the old me or a new me?

I sleep so easy on the ground and marvel how well my body responds to daily stress. Except for the bout of chafing, no unusual aches or pains and my left knee hardly clicks at all. I must be doing something right. Or maybe, like the sneezy, runny-nosed allergic kid I was, moving keeps me feeling better.

"Patti?" Barry, outside my tent. "It's 5:30."

"Yep, I've been awake a while." My bladder nudges, *he's up!*

"You said last night you wanted to get a photo of us before we leave."

I roust to take a selfie, then set about my morning bike check (after I hit the head).

- Tire pressure—100 psi front, ninety back, right on. Top-off needed about every three days.
- Spin wheels—true.
- Visual check on rack bolts, water bottle cages, fairing mount—secure.
- Rain yesterday—lube and wipe the chain, a few drops on derailleur pivot points, rear pulleys, and pedal clip mechanisms.

Good to go.

As Jason recommended, I sail south out of town on Main Street. "Turn west at 34. It'll put you right back on track." A glorious seven-mile, tailwind ride (past another beehive drop-off) ends when I turn due west. The bike leans into a nasty side wind.

"Now my tires will wear on the right side," I mutter.

The blast is incessant. I take to greeting cows as I inch by. "Good morning! Good morning!" One white-faced cow gapes at me. "I'm thinking about having a burger for lunch. How 'bout you?"

No reply.

A black calf startles and runs neck and neck with me. Instead of taking refuge behind an adult cow, he races past. She startles and runs too. And the next and the next, a wave of runners. The herd stretches out for a good quarter mile, I fear a stampede.

In a blink, the wave of cows stops.

Forty miles later, just before noon, I dip into Napoleon and drool at the sight of a Super Valu Supermarket. At a shaded bench in front of the store (not quite out of the gales), I devour two flour tortillas loaded with a seven-layer-bean-dip, drain a thirty-two-ounce Gatorade, and polish off a cherry yogurt.

Fortified, I face the now-turned-west wind. Hills or no hills, it doesn't matter, I limp along in the granny gear at five to six miles per hour.

I belch another mutter. "At least a straight on headwind makes the bike easier to control."

The road snakes between two peculiar little lakes, not unlike the one yesterday divided by an abandoned road. Will this road suffer a similar fate? A row of dead, bleached white trees rise from now-brown, now-green waters; two peeling clapboard houses lean west several yards offshore, whitecaps breaking at the windowsills. I half expect to see a billboard advertising "Surf North Dakota."

A billboard of any kind would be welcome, especially if it gave privacy for a bathroom break. This open land is a challenge. At a sweeping curve I lean my rig against a road marker and take what cover I can behind brush on the opposite side of the road. Curious eyes of passing drivers should look to my recumbent instead of me.

A rumble swells like the thunder of an approaching train. By the time I scurry back, a quarter mile of motorcycles reverberates a primeval pulse through my core. I am rooted.

Two by two the riders roar by, one by one they drop a left fist in salute. A brotherhood on two wheels.

Damn the wind.

I push on to my night's stop in Hazelton. The image of all those motorcyclists cruising through this town of 235 people brings a smile. They might have doubled the population.

Tucked a few blocks from the town proper (you can hardly call it a downtown) is the Hazelton City Park. Three RVs park on gravel sites with electrical posts. A sand-filled square of playground equipment sits in the middle of a half-a-football-field sized grassy area; between the playground and a bathroom building is a pavilion with a few picnic tables. Except for a pre-teen girl and a younger boy, both coloring in the pavilion, no one else is around.

I take refuge next to a barrier of bushes. Beyond the bushes are two makeshift tents of cardboard boxes and tarps, as if a homeless person constructed them. With my little tent and gypsy-like gear, I fit right in.

In the shower, I find another tick, this one stuck on my right ankle. It is easily disengaged. And dismissed. So nonchalant about a tick. *Who am I becoming?*

Yet I am wind weary. Mom senses it when we FaceTime. "You should rent a van and come home."

"I will get to Missoula anyway, although I admit I lean toward taking the train home."

"Good."

STATS
Wind gusts to 45! 66.89 miles, max speed 25, ride time 7:31, avg 8.9, TOTAL 1073.68

18

Gone With the Damn Wind

Sunday, June 26, 2016
Hazelton to Bismarck, North Dakota

Boundaries meant to keep the youngster safe only tempted her to explore. Allowed to ride the barely small-enough Shelby Flyer around the square city block from home, she was not allowed to cross any streets.

Head down, tail up the girl turned left at the first corner and dismounted at the next. Heart racing, she ran her steed across the prohibited street; the front tire smacked the curb, the bike reared, and the handlebar met bottom teeth through her lip.

Blood spewed.

Breathless, the girl retreated, pain not yet registered. She pedaled the counterclockwise loop home to confirm a concocted story. Two houses away, the girl spotted her mother sitting on the front porch. Pain and tears erupted.

"Patti, what happened?" Her mother leapt to catch the girl, now a sobbing, wet, red mess.

"I…was riding around the block…you know where that big tree pushed the sidewalk up? My head was down…the bike bounced up and hit me!"

Her mother didn't let on if she believed the girl's story or not, she just hustled her off for stitches.

Thankful for Sunday morning traffic and a six-foot shoulder on Highway 83 out of Hazelton, I am wind sick, perched on an ocean-going dingy buffeted by squalls. Silence breaks in brief interludes—I hear my lungs suck in and expel. A gust bellows and thrusts me toward the ditch. Moments later, I hear crickets chirping. For a quarter mile, a purple field blurs my vision. *What is it? Clover?* A heady scent reminds me of the lavender bush in Andy's backyard garden.

A flurry of sparrow-sized birds, black with orange heads, tumble, recover, and soar ahead, my morning escort. In the trough of a road-wave, starling-like birds swoop to circle my passage. There. A lake glistens, maybe a half-mile east. A brown sign rebuts in white letters: "Long Lake National Wildlife Refuge. 2 miles."

I am losing the ability to judge distances.

At Moffitt, a town of no population with no services, I turn west, headfirst into a blast furnace. Four club riders heading east barely have time to smile and wave before the wind carries them away like an illusion. My mouth opens to wish them "good luck if you have to return to Bismarck," but wind smothers my words.

I am back on the tandem with Lou, in France nearing the coast during PBP. Tom, a top American rider caught us. He started with the last and fastest group. (Lou and I started with the first wave at 4:00 a.m.) After poor luck of several flats right out of Paris, the lead pack dropped him. "I've given up on fighting for the win."

Tom rode a few kilometers with us. "Boy, you can really see a lot when you ride this slow." And off he went in a Road Runner puff of smoke, leaving Lou and I humbled. We thought we were riding strong.

Tom was right, there's much to see creeping along. With the wind not trashing me around from the side, I don't have to concentrate on keeping my dingy straight. The prairie is vast compared to France. Jumbles of boulders the size of F150s and small Winnebagos dot panoramic fields. Amidst a pile of broken concrete and rusty farm equipment, a deer observes me. At a glance, it vanishes, like Tom's rear end in the French countryside.

Half-way up a mile-long climb, I disembark, unable to continue. The bike rolls backwards. I lean into the handlebars and force a step. This North Dakota wind is NOT a normal wind, it is a wicked-witch-of-the-west wind coming for my little dog Toto.

Lyrics from "The Dream Before," another Laurie Anderson song from her album *Strange Angels*, filters through the still-constant beat of "Ramon." A story about history repeating. Winds of progress blow an angel to the future, instead of to the past where she wants to fix things.

Is my ride a journey back? Or forward? Did Forrest Gump stop running because he grew tired of fighting his-story?

Continue I must.

I rest at least six times before cresting the hill. A cloud shadow swallows the desolate landscape. The wicked-witch-wind reads my mind when I am about to shift to a harder gear and whaps me broadside. I concede the gear and spin as I can. There is no escape, no coasting downhill, only pedaling keeps me moving.

A billboard looms lonely out of nowhere at a sharp zag in the road.

ABORTION STOPS A BEATING HEART

The monster-sized words punch a line drive into my gut. I tear my eyes away. Anger displaces my breath. Who posts such a hideous message to shame others? What about our hearts?

I am relieved when the road zigs back into the headwind and I can fight the adversary I cannot see but feel. "Let me pass!"

The wind. It just is.

No shoulder to cry on, the road edge drops from the white line at forty-five degrees into a culvert that stretches seventy-five feet to power lines. Flowering milkweed in the ditch look like aliens birthing; lowly little sunflowers turn their faces from the wind; other tiny white blooms that look like paper poppies sneak along cracks into the roadway. *What are they?*

"Come here my deary," they call.

Now the milkweed gathers too, in bunches at the edge of the road, straining to reach me. I am glad these body-snatching plants are rooted, unlike me who can steal away at five miles per hour.

Keep pedaling or lose my mind.

Two hours to ride eleven and a half miles, three to reach twenty-six. An incredible average of 8.8 miles per hour, seventeen miles to go to General Sibley Park in Bismarck.

I need to eat.

Sanctuary is a sliver of shade on the leeward side of a manufactured home turned schoolhouse. I inflate my Thermarest and settle in for an eerie picnic next to a deserted, sandy playground. Sun glints against a polished slide. *Are ghost children wavering on that metal swing set?*

Sliced avocado and tomatoes wrapped in a sun-warmed flour tortilla taste like a summer harvest and sooth my prickles. Despite the acid setting my blistered lips ablaze, I eat two.

Off route, and some four miles south of Bismarck along the Missouri River basin, General Sibley Park is a welcome oasis. My choice of tent sites in a tree-shaded field away from rows of RVs.

Wind-exhausted, I stumble about setting up my tent. A cyclist hauling a trailer pulls up to say hello. "I did over 100 miles today." By Chuck's wide grin I can tell the wind blew him in.

I scowl. "I'm glad someone enjoyed the wind. Did you see a grocery store on your way in?"

"Dan's Market, five miles back into town. You can take a bike path almost all the way. I stopped there but didn't feel comfortable leaving my bike outside."

My shoulders sag. "I'm not going there now, I'm tired of fighting the wind. I'll stop on my way out."

"You should think about taking 94, it has a wide shoulder and is less hilly."

Grateful for the tip, I tell him if he continues on 94 he will miss the Honey Hub in Gackle. "Hey, do you want to share this site?" The space is large enough for a group of twenty-five riders. Although the $12 tent fee is not exorbitant, half the cost saves money.

"Thanks," Chuck said, "but I don't want to intrude."

Does my wind-weary, sour self deter him? Fair enough. But I am surprised by his fear at the grocery store. On the road for over two weeks, no one has ever bothered my bike or campsite. I don't even carry a lock.

Four yellow-shafted Northern Flickers spatter about a gnarly cottonwood tree less than twenty feet away. Two of brighter colors peck from a deep hole at the base of the tree and feed the others. A family? They flit and fight and mount each other and squawk with abandon. My tired presence is of no concern.

I am spellbound.

Is it the birds that flick away the strain of the day? Or the email exchange with my Rover-assistant, Debbie?

"Do you need anything?" Generosity is this woman's middle name.

She gifted $300 to my trip and wants to do more.

"Can't think of anything." Except maybe a shoulder to whine on. I share my indecisions about the trip back. Train or not? "I change my mind three times a day."

"Let me know if you decide to take the train. I can help."

She so wants to be part of this adventure. Some years back we took a three-day, self-supported tour of northeastern Michigan. While Debbie has ridden many supported bicycle tours, carrying gear on her own was a first. "I had no idea it could be this much fun."

Calm settles me like a cool breeze. I try to FaceTime Mom but it won't connect. I call instead.

"I shut off my iPad." She sounds depressed. "A lady here fell in her bathroom and cracked her head. It was a heart attack. She ended up dying." She tells me of two others being treated for cancer. "It makes me want to run away, but I don't know where!"

"You could come on a train ride with me."

Her voice perks. "That would be fun." She really wants me to come home on the train, even though she knows I'd have to ship my bike separately. "You could just sell your bike and buy a new one when you get home."

No, I couldn't.

STATS

43.75 miles, max speed 25, ride time 5:10, avg sped 8.4, TOTAL 1117.52

19

Chance Meetings

Monday, June 27, 2016
Bismarck to Glen Ullin, North Dakota

Switch hands? Not happening, in spite of what the nuns wanted. True, the girl learned to throw and catch right-handed as there were only right-handed mitts in the house. But writing? No way.

"Then you'll have to balance this eraser on your wrist. We can't have smudged ink."

The girl rejected the eraser and learned her own way not to smudge.

Patti's outdoor activity rule of dress: If you are warm enough when you start, you are over-dressed; take a layer off and start out chilly; in ten minutes, you'll be fine. Ignoring my own rule, I wear fleece for the ride to the supermarket. Yep, over-dressed. On the sidewalk-that-doubles-as-a-bike-path (glad to not be fighting Bismarck's morning rush on the road), I stop to stash it in my pannier.

Dan's Market, as Chuck named it, is really Dan's SuperMARKET. With a parking lot almost twice the size of the store, the store dominates a busy corner near where the ACA route turns west. I don't share Chuck's fear at leaving my bike unattended. The long rig might intimidate an opportunist from hopping on and riding off, and good luck lifting it, you'd need a truck to haul it away.

Fifty-five dollars later I drag four grocery bags out of the too-many-choices store. To be fair, a few things are not food—sunscreen, bug spray, Blistex, Aleve PMs, and toilet paper.

A man walks past and grins. "Storage problems?"

"Plan to eat some right here." I cut our conversation short when I see my yellow rain jacket in a bunch on my seat, instead of strapped across my gear. *Did someone mess with my stuff after all?* Wait. Here's a note with a phone number.

This dropped off two blocks south. Thought you might need it. Howard.

I glance around. No one is paying attention. With intentions of calling Howard later, I tuck the note into my map case.

Ha! How 'bout that, Chuck?

My smirk floats me through Bismarck and over the Missouri River on a wide asphalt surface that is now an actual bike-path. A man in a white SUV beeps his horn. Howard? I wave. Nope, he doesn't slow.

When the path ends beyond Mandan, I take Chuck's suggestion and jump onto I-94, which parallels the ACA route through to Montana. The cattle-guard on the entrance ramp is evidence that even though this ribbon of highway connects me to Michigan, I am a long way from home.

The shoulder is as wide as a lane and rumble strips give a modest sense of security. Crossing exit and entrance ramps is tricky, but years of playing in traffic serve me well. The biggest challenge is avoiding shards of indistinguishable metal or bits of wire from exploded retreads.

A dual-wheeled pickup truck pulling a fifth-wheel as huge as the skyline rumbles by, the brand name Montana plastered across the back. "Montana!" Almost every other RV is a Montana. Getting closer.

Gradual climb after gradual climb, wind from ahead, wind bursts from the side when a truck passes, the roar of tires like ocean surf, then silence when no trucks come. A mesmerizing pedal. Until I sense a softening. A flat? Not surprising if true. My front tire looks fine, can't get a good view of the rear without swerving. The map shows a rest area beyond New Salem, I'll check the tire there.

Cement picnic tables at the rest area nestle against three-sided brick shelters. Shade and respite from the wind, what luck. I roll into the nearest one.

The rear tire squeezes hard, but I want to check it with the gauge. It's packed away in the tool bag on the seat back, blocked by gear. I fumble, the bike slides, my leg catches, and it's a scuffling match to right it.

An East Indian-looking man, wearing dark sunglasses and a faded American-flag t-shirt, rushes up. "You need help?"

"No thanks, I've got it. Just checking my tire. I thought I was getting a flat, but I think it's okay."

The man asks me typical questions in simple sentences, English is not his first language. In return, my voice sounds deep and distant.

"Be safe." He takes leave to join a woman wrapped in a horizon-blue dress and four girls in bright pink and blue shirts at another table. I set to checking tire pressure. Ninety-five pounds good.

My head floats. Am I bonking again? Is the heat getting to me? Food. Yes.

The man returns to find me rifling through my feed bag. "Come share our meal. We have chicken. Our kind, you might not like, but you welcome." He sweeps his arm with a half-bow, indicating his family behind him.

"Thank you." *Should I bow too?*

The rest stop is busy with travelers and truck drivers. A tanned man wearing black jeans, cowboy boots, and a white Stetson hat strides across the grass and glances over.

No longer connected to North Dakota, to the United States, to the Western world, I float above the earth, a sojourner in distant lands. I imagine what this man sees—a ragamuffin lady (he isn't sure it's a lady) in clothes too big, joining dark-skinned foreigners for lunch. Does he wish us to go home? Or is he curious to join us? I suspect I have much yet to discover.

My host's voice draws me back to earth. "Tell my girls what you do."

Wide-eyed, the girls smile and nod at my adventure. From cloth bags, the wife unloads paper plates and towels, a glass bowl, round plastic container, naan bread, small Ziplock container with lime slices, bottles of water and Coke.

I finish my tale. "My name is patti."

The man tells me his name, but I cannot understand him. He repeats it several times. I ask him to write it in my small riding notebook. He gestures to his oldest daughter. She writes, "Santosh Ramdam."

Santosh's wife hands me the glass bowl filled with orange chicken legs. The aroma of cumin and exotic spices wash over like the smoke of sweet grass. I take one and pass the bowl on. The wife pops the cover off the plastic container. Filled with rice that doesn't look to be enough for all, I pass and offer what I have—flour tortillas, a hunk of melting cheddar cheese, and a banana.

All hands wave "no."

My first ginger bite of the spicy chicken floods with flavor, then a firestorm blazes on my scabbed-over lips. Santosh chuckles when I dab with my ever-present kerchief. Bites alternate with dabs.

Santosh offers, "More." Torn, I fan a "no." This chicken is delectable. But. My. Lips. Santosh might mistake my hesitance for distaste. How can one survive something at once so achingly wonderful and so painful?

The drumsticks and rice disappear without silverware. I tell the wife how delicious the chicken tastes. She smiles. Santosh says she does not speak English.

I ask, "Where are you traveling?"

"We go to Canada, to our people." He says he's traveled much of the United States. "Is a big country, it never ends." In a split turn of conversation, he asks, "Do you believe in Jesus?"

Without hesitation, "Yes." I believe a man lived. Here, with my new friends, I need not dismiss his virgin birth and heavenly ascension.

"You go to church?"

"Not anymore."

He nods. "Ah."

I sweep my hands to the horizon. "This is my church." The world, the countryside, nature all around us.

"Interesting." He leans over the table, eyes glaze in deep thought, as if not seeing me. At last he says, "You are on journey."

"Yes. Riding to Missoula."

"No, your life, a journey. You are…." His words stumble. "You. Focused." A pause. "I get energy from you, your…power."

"I don't understand."

He struggles, repeats what he said, unable to find words to explain what he means. "You have children?"

"No. My husband does."

"But, family?"

"Oh yes, a large family."

He thumps his palm on the table. "You must write down, to share with family. Your journey."

What? How could he know I've been writing, writing, writing, posting on Facebook, feeling so encouraged by positive reactions to my posts that I've decided this, this "journey" is to be my book. Finally, I have a venue, a vehicle, I might have something to say, that my life is worth telling, even if it's just for me. It is what I need to do. What does this stranger see in me? Does he perceive my journey as the vision quest I sense it is? The miles I travel are more than nostalgic, as if I must learn or re-learn something. Or un-learn.

Santosh leans back. "You give me much energy. Can I give you t-shirt?"

"Sure."

He talks to his wife in their language. She and the girls head to their car with the packed-up picnic and return with a green shirt. On the front is a cross with the words "Nepal Baptist Church."

"It is huge!" I blurt holding it against my shrinking body. Size medium. "Do you have a small?"

Santosh speaks to his children. The shirt they bring now looks too small, but I am not about to renege.

Each one gives genuine hugs "goodbye." Santosh palms me a $20 bill.

"I don't need this."

He backs away with a smile. "Dinner from us."

The hot, windy miles remaining to night's rest in Glen Ullin are effortless.

West of town, population 802, is the Glen Ullin Memorial City Park, an enchanting green among vast, open ranges. A rust-red dirt drive leads me past a pit toilet, an old rail car perched on a length of track, and 100 yards of not-very-tall trees. I stop at a bulletin board with instructions on self-registering when an older woman takes cautious steps from a nearby trailer. The camp host.

"Take any spot you like along the grass there." She points to either side of the drive. Beyond her trailer are sites for RVs and campers. "Six dollars for a tent. The WiFi password is on the board, and over there we have new water spigots."

Dirt scars in the thick grass show where water lines were dug. Mounted on four by fours (with electric outlets), the spigots line a tall hedgerow, protection from the ever-present wind. Perfect.

I call Howard to thank him for rescuing my jacket. He says, "I was traveling south and saw it fall off your bike. Was going to yell out my window but was past you. I saw you cross the road and head to the grocery store. When I came back the jacket was still there, so I stopped and picked it up. I looked for you in the store but couldn't find you, so I left the note."

"Yesterday I met a cyclist who worried about leaving his bike at the store. Sure wish he saw what you did. You really made my day."

Howard's smile beams in my ear.

Evening brings calm. I sit at a wooden picnic table to write about this amazing day. I hear the crack, crack, crack of some birds. My pen pauses. A few feet away, brown-headed cowbirds peck at something in the grass. Two more flit to join them. Insects? I can't tell.

A myriad of noises tease for my attention. Eyes close. Listen. Cows moan, trains groan. Trucks air-break coming into town from the south, a motocross bike revs back and forth in the north. Someone shoots off firecrackers.

When I open my eyes, a poofy, gray and black and white bunny, with a white ring around its eye and neck, rests on its jumping feet only a hair's breath away. I shift. The bunny springs eight feet away. Pen in mid-air, I freeze without eye contact. The bunny hops toward my tent, nose scouting at warp speed, and stops. Another hop and the bunny will enter the open vestibule of my tent.

Background noises drift away; it's only me and Bunny.

When I call Mom, she hears my excitement for the day. I email her a photo of Santosh and his family.

"I love it!" she says in her email reply.

"I feel like someone is watching over me." I send back.

She emails one word, "Yep!"

Andy emails too. He received my "fourteen-day" letter. "I'm impressed with how introspective you are." So sweet.

Today's chance meetings fill my mind. Coat rescue by Howard. Philosophical lunch discussions with Santosh. How Bunny lingered.

What is happening to me? Have I ridden through a Chrono-Synclastic Infundibulum in the vastness of these wind-blasted plains? "A dangerous thing to do," Kurt Vonnegut writes in his book, *The Sirens of Titan,* a quantum physics kind of place where all truths are equally true, a place where time curves "toward the same side in all directions." Vonnegut likens it to an unpeeled orange.

In *Between Time and Timbuktu,* a 1972 teleplay based on Vonnegut's stories, poet Stoney Stevenson wrote a prize-winning jingle to win a trip through a Chrono-Synclastic Infundibulum between Earth and Mars. As I remember it, the ending shows Stevenson pulling himself up from the dirt of a fresh grave. He brushes off, looks back at the stone marker, and skips away whistling the song, "Pack Up Your Troubles in Your Old Kit-bag and Smile, Smile, Smile." The camera pans to the gravestone.

"Everything was beautiful, nothing hurt."

STATS
62.05 miles, max speed 34, ride time 5:29, avg speed 11.3, TOTAL MILEAGE 1179.60

20

And Here I Am

Tuesday, June 28, 2016
Glen Ullin to Belfield, North Dakota

"Oh, I can't let you check these out today," the librarian scolded the girl. "You never returned 'The Black Stallion.'"

"I did so!"

"We've no record of it. You must pay for it. With late charges that will be $28."

The girl was pissed. Of course she returned it. Every Saturday she walked four blocks to the library branch, returned the four books she checked out the week before, and checked out four more. Nothing she said convinced the librarian. She dropped her selections on the counter, trudged home to face her mother, and returned when she saved enough baby-sitting money to pay.

Months later, checking to see if 'The Black Stallion" reappeared in its place on the shelf, she found it—the exact book she checked out and returned, her name and due date on the book card in its front cover pocket. She stomped to the librarian's desk.

"Look what I found, the book you said I stole."

The librarian didn't even look up. "Not my problem."

Beside herself, it wasn't the money that irked the girl, it was the fact the librarian didn't believe her.

Sometimes my beautiful ride does hurt. Not in my hands, shoulders, or neck, the recumbent affords a pain-free, unobstructed view of the world. (No more upright touring bikes for me.) But fighting these unearthly winds have tuned my leg muscles tighter than over-strung guitar strings. Two Aleve PMs before bed helped. Still, my bladder wakes me at 5:30 a.m.

Fishing for the headlamp I keep stashed in a tent pocket (with my glasses) I snag a snapped metal frame missing one lens. Damn. I hoped this old pair would last the summer. Good thing today's ride takes me through the city of Dickinson. Found a laundromat with my iPhone last night; now I need an eyeglass store. Forty years ago, I'd search for a phone booth with Yellow Pages hanging on a wire.

For now, at a chilly and damp forty-five degrees, my bladder can wait. I sink deeper into my sleeping bag, thoughts fly to France, when Lou and I were half-way through PBP, leaving the coastal town of Brest.

Tired clumsy, we fumbled with kicking into our toe clips (the old-fashioned kind with straps) on a slight uphill leaving the checkpoint. Off to the side, a bent-over old woman wrapped in a shawl gripped the hand of a six-or-seven-years-old little girl. The girl waved in a frenzy. We nodded and put our attention back to our pedals.

The woman yelled in French and the strength of her voice planted our feet. The girl dragged the still-babbling woman to us. We did not understand. The woman huffed, grabbed my shoulders, and pulled me to her face. She kissed me, first on one cheek, then the other. She gestured to the girl who jumped up and down clapping.

"Léopards Roses! Léopards Roses!"

I bent over and the girl planted two rose-petal kisses on my cheeks. She did the same to Lou.

Those kisses propelled Lou and I up a long, straight climb and on to be the first women's tandem team to complete the event. I hope we inspired the girl to follow her dreams, to know she can do anything she sets her mind to, that she is stronger than she can imagine.

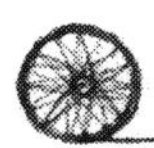

And here I am at age sixty, riding my bicycle once again. Relearning I am stronger than I imagine. That life is for living dreams. That death is the worst that can happen and there is no getting out alive.

Will this ride inspire others?

Like Forrest Gump's fans who started running behind him. With Bob Seger's "Running Against the Wind" playing in the background,

Gump ran through a pile of dog poo without missing a step. A follower questioned him. "Shit happens," Gump shrugged. In the movie, the man made a fortune on bumper stickers.

Dad loved that saying.

The memory of the French girl's sweet kisses warms my cheek even as my bladder screams, "Let's move!"

After the long walk to the outhouse, I notice Bunny hanging at my tent again. Like a favorite stuffed animal, except for the twitching nose, he is motionless as I pass a heartbeat away.

I glance away and he is gone. *Did I really see him?*

If I did, or didn't, Bunny brings me luck with a slight tailwind. Dickinson before noon, twenty-three miles in an hour and a half, I expect suppertime in Belfield. And Montana tomorrow!

Eyewear Concepts is on the same busy road as the laundromat, a few blocks north. Alissa, a young and very pregnant optician's assistant, inspects my frames. "We can solder it; not sure it will hold. It'll be $50 for the repair, but our technician is out to lunch."

"What about reading glasses?" I need some way to read the maps. My sunglasses are a distance prescription, not bifocals.

Alissa checks my bifocal script and shows me $20 reading glasses. Another suggestion, "We might have a frame that would fit your lenses."

She finds a child-sized frame to fit my wee face. "These could work, they are only $69. I'll have to file the lenses to fit and glue them in so they'll stay."

I can't decide. "Tell you what, I'll go do my laundry and think about it. Maybe your tech will be back by then." Lunch too, will aid in my decision.

The technician confirms Alissa's doubts. "I can't guarantee a weld will hold."

New kid's frame it is. I am back on I-94.

Funny how the faster I go, the more impatient I become. Miles drag. Gradual descents break long climbs, most of which don't require the granny gear. The ACA maps switch from the Northern Tier Route to the Lewis & Clark Trail and now include a terrain profile. Terrified to look at it, I don't need the profile to tell me I've been gaining altitude from

Fargo. I feel it in my legs.

On one long, long grade, my mind wanders. There is something I regret.

Not too many weeks before he passed, Dad woke up on the hospital bed next to Mom's new single bed in their senior apartment. Soaked through to the sheets, despite wearing Depends, he couldn't do anything about it for himself. He was cussing angry. Mom dialed the home health team for help.

We waited. The aides had fifteen minutes to respond. The clock ticked beyond.

"This sucks," Dad swore.

"Hang on Dad, they'll be here soon."

Tears glinted on his eyelashes like raindrops. He turned his head toward the closet, he turned it away to face the window, he stared into the air above my head. He grabbed the side rail, knuckles whitening. Through pursed lips, "I'm lying here in my own piss!"

Why didn't I get him up myself? While it would have been hard to lift him from bed, into the wheelchair, and then the bathroom, I had transported him before. Well, that was when he could still help support himself. I guess I didn't think the home care team would take so long. Mom paid a lot of money for their help, the extra service was one reason for moving to this senior apartment.

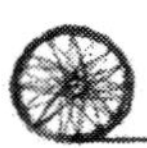

I fight a steeper climb and tears stream. "I am so sorry Dad." I regret not easing some of his hurt that morning, not taking care of him myself, making him wait. By the time I drop my chain into the granny I swear I hear Dad whisper, "It's okay, Pat."

As I inch along, the forty-five-mile-per-hour-wind-gust day into Bismarck comes to mind, with the ABORTION STOPS A BEATING HEART billboards. Yes, there were two sucker punches delivered anonymously around that curve, with no chance for my heart to press back. I *am* sad about my long-ago abortion. Who wouldn't be? But I don't berate myself. Should'a, could'a doesn't change things. I should have, could have been more careful (and was afterwards, thanks Dad for that "don't be sorry, just don't do it again" lesson). I don't regret my decision. Who knows how my life would have turned out otherwise? Would I have learned massage at the right time to help Liz die? Or the other hospice patients I touched? What about Mom after her bypass surgery? To this day she claims the massages boosted her recovery.

No. I can't regret it. With my sex education a "wait until you get married and practice rhythm," how could I not fail when I rejected Catholic standards to follow my body's desires? I am sorry for the mistake of getting pregnant, but if a god condemns me to hell for being human, he isn't a god I want to love.

I crest the climb and grimace the front shifter out of the granny. A new time zone matches my zoned mind. I am now two hours behind Andy and Mom.

Before bed, Debbie sends me an email, again wanting to help me keep going. Her words echo. "You are on the Warrior's Path. You are the closest person to a warrior that I've ever met."

I need nothing.

POSTCARD FROM THE ROAD 6/28/16

After a seventy-mile day, I wheel my bike across a red dirt drive into a scrufty campground behind a hotel/restaurant complex. It is nothing more than a mowed field with a few scrawny trees.

Site #29 is mine, complete with a tree, picnic table, water spigot, and an electrical box (no need to steal electricity tonight). I am one empty site away from the shower building, which is unlocked with a credit-card-like key. Planted on the other side is a large, fifth-wheel camper, one tire completely shredded to the rim. A few more decrepit rigs like this dot the rest of the grounds.

After a spicy southwest steak, bean, and rice meal, I sit at the picnic table writing. A dusty black, extended cab, four-wheel drive pickup truck pulls in facing me. A young man dressed in dirty green company work pants and shirt jumps down and says with a drawl, "Hi neighbor!" He strolls over and takes a seat at the table.

Brad has been working in the oil fields in the next town for a few weeks. He's from western Oklahoma. "It's a long *way from home," he laments. He says he usually works four or five hours from home in Oklahoma, but there isn't work available. "So the company sent me up here." He might be here for three months. Brad misses his sons, ages four and ten. "But they have a good life, that's what matters," he says.*

The campgrounds in Fargo, Hazelton, and Bismarck had similar campers staged for workers. Young men would straggle in at dusk driving pickups or work trucks. They went to bed early and were gone before I'd break camp.

"You have to provide," Brad says. "And there isn't much work in western Oklahoma."

STATS
70 miles, max speed 32, ride time, 5:31, avg speed 12.6, TOTAL 1249.81

21

Welcome to Montana

Wednesday, June 29, 2016
Belfield, North Dakota to Wibaux, Montana

"Do you have a twin brother?" the first male teacher at Immaculate Heart of Mary (IHM) grade school asked the girl.

"Why?"

"I was driving home, and I saw a boy riding a bike down the sidewalk. He looked just like you."

"Oh," the girl smirked, looking down. She couldn't help herself. "Yes. That's my twin brother, Steve. Something is wrong with him; they won't let him come to school."

"You're making good time!" A man yells at an I-94 rest stop overlooking the Painted Canyon.

"I guess I am today. How do you know?"

"We passed you."

I can't help but notice that one of the two bicycles stashed in the back of his hatch-open SUV is a titanium road bike. This guy is not a casual rider. He grabs my hand when I say I rode B'76 and am headed to the anniversary celebration.

"Are you going through Helena on your way to Missoula? That's our home. We'd love to put you up if you do." Bruce and Susan Newell, with their daughter and her friend, are heading home from a family wedding in Minnesota. They brought bikes to ride while there. Bruce and Susan did a west to east transcontinental crossing a couple years ago.

"Not on the way there, but I might if I ride home."

"It's a great ride from Missoula to Helena. You won't have any trouble."

Bruce and I share emails. "You should get off 94 after Medora. Old Highway 10 is a spectacular ride. It goes through country like this, but it's up close."

This rugged tapestry of red and tan striations, marking the eastern edge of the Theodore Roosevelt National Park and the vast North Dakota Badlands, reminds me of a request my boss made before I left home.

"Bruce," I say, "will you take a photo of me by the canyon? I'm the editor of a local paper back home and my boss wants me to write an article about my trip. He thinks I'll be riding past the Grand Canyon!"

"That would be the long way to Missoula," Bruce says. We all laugh and head to the overlook. "Holy *wow*!" He exclaims when I hand him my Nikon D7100. "You're carrying this?"

I know, it's an extravagance I haven't used enough to warrant the weight. Bruce fires off a volley. A passerby offers to take a shot of me with my new friends.

With renewed energy, I cruise down the entry ramp and hear boisterous barking. The side of the highway teems with…what? Prairie dogs? The squirrel-sized critters stand on hind legs and blend in with the dry brush; heads twist back and forth like signal flags; they bark and scuttle between holes in the sand, diving in headfirst only to pop right out again. They are amusing, but now I must pay attention to staying on the shoulder. Traffic is picking up.

Past Medora, at the top of the exit ramp to Old Highway 10, I double-check the map, not sure I want to leave the easier highway grades. A road worker in a white pickup truck asks if I need help.

"You'll do best to take 10," he says. "We're doing construction ahead. If you stay on 94, you'll be okay if you keep to this side. Just watch out for our trucks. But 10 is nice."

His words cement my decision. "Thanks for the advice, I'll take 10."

Old 10 drops me like a roller coaster, bluffs close in at every curve just as Bruce stated. Gradual descents and inclines no more, the road plunges into and out of every gully. I struggle and resign to walking the steepest climbs.

Am I bonking again? The fierce landscape scratches my mind. In the endless sky ahead, darkness looms. *When will the storm hit?*

I press on.

Sixteen miles later, hoping for refuge at the convenience store in Sentinel

Butte promised on the ACA map, there is only a run-down gas station. An extended roof sags over a single gas pump that looks as if it has stood there since the 1940s. It appears vacant.

Ah, well. I lean my bike against one of several buffalo-sized boulders around a flagpole and a wooden sign honoring Custer's march. Before I finish extracting lunch from my food bag, quarter-sized raindrops thump my helmet and fairing. I race my rig under the station's overhang. Movement behind a murky window catches my eye. One of two men sitting at a massive round table sweeps his arm in welcome. "Come on in," he mouths. Not vacant after all. I gather my tortilla-wrapped hunk of melted cheddar cheese, a handful of beef jerky, and the last of the Gatorade I picked up in Medora. It takes all my body weight against the heavy wood door to enter.

"Take a seat!" Sun-bleached blond hair and a graying mustache frame sparkling gray eyes, the man's smile is canyon-wide. He wears a University of Montana Grizzlies' t-shirt over dusty jeans. Rick is the station owner.

"I went to school in Missoula. We get lots of bikers through here. Help yourself to a muffin. Care for some banana bread?" He slides the loaf over and points to a dish of muffins and a cookie sheet filled with white chocolate candies. "My son got married over the weekend and I need to get rid of this stuff."

"Thanks." I grab a muffin.

The other man, Boyd, sets a half-eaten sandwich down and gets up to refill his coffee. "Care for a cup?"

"Sure." Coffee and a moist muffin. *Mmm mmm.* Nice place to sit out the now-pelting rain.

A young man in bib overalls and a sweat-stained John Deere hat comes in and constructs a pile of nacho chips and cheese from a back-counter dispenser. Another man stomps his feet on the rug at the door, grabs a sandwich from the cooler, microwaves it, and joins us.

No wonder Rick has such a huge table.

When an SUV pulls in at the pump, Rick runs out to pump gas. *What year is this?* Stranger in a strange land, am I a time-traveler too?

An older woman, well, maybe a woman my age, comes in with a little girl about four years old. "Were you out haying?" she asks the man in the John Deere hat, sits, and hoists the girl onto her lap.

"Yep, the rain chased me in," he says between mouthfuls.

Boyd offers the cookie sheet to the girl who has been eyeing it. She looks at the lady for the go-ahead. With her nod, the girl launches for a candy.

The woman picked up her granddaughter from a summer day camp in Beach, the next town over. I must reach Beach before its post office closes to pick up an allergy prescription refill Andy sent.

"You'll have a nice hill to climb before you get there," the woman reports. The men around the table agree. I hope that, like directions given by most locals, they are wrong.

Rick returns and picks up a clipboard from next to an old-fashioned mechanical cash register. "That was seventeen gallons." He writes the amount and her name on a chart.

"Thanks," the woman answers. No money exchanged. "Make up a dozen donuts for me, will you?"

Rick sets about boxing donuts, spreading a thick layer of maple glaze over each one. Gas and custom-iced donuts. Be careful what you assume with first impressions. Rick's ramshackle station is a hub of commerce and connectivity. To think I almost missed it.

Sunlight seeps through Rick's window, I turn to look out. A gray, shaved cat sits outside on the sill and doesn't budge when a woman cyclist rolls up. The woman parks her laden bike on its kickstand next to a metal bike rack I didn't see when I arrived. She is soaked. After stripping off her neon yellow jacket, she hangs it and her helmet from her handlebars and joins us. From Colorado, she is riding from the west coast to the east with her neighbor. The pair hooked up with a couple on a tandem pulling a B.O.B. Trailer, a recumbent trike rider, and three others, all behind her today.

"Too many people," she laments. Can't say I blame her. There is something special about traveling alone. I beg my departure before her group arrives.

Smooth, new pavement from Sentinel Butte dries quick. The locals are not wrong. Between me and Beach, the hill looms as a castle wall, a square cement silo launches like a turret at its peak. I am determined, renewed by the positive energy in Rick's station. As I hammer up and away, the tandem and recumbent scream down and salute a thumbs up.

I return a wave and keep pedaling. And pedaling.

There is joy in Whoville when I crest and let gravity capture my fall. Sunglasses cannot keep the wind from tearing my eyes. In the distance, an ant crawls toward me. A second later the ant is a cyclist standing on his pedals, my words of encouragement lost as I zoom by.

God, I love riding my bike.

A minivan slides up next to me at a stop sign in Beach and the woman driving it waves. I gesture to roll down the window.

"Do you know where the post office is?"

"No, I'm just passing through too. I'm sagging for a group heading east. Seen them?" The woman carries her riders' gear and follows them in case they need help.

"I passed them on my way from Sentinel Butte." She makes a quick u-turn.

Across the intersection, in front of a small, flat-roofed building that looks like a machine shop, three men cut boards with a circular saw. I coast over for directions.

"Go back three streets and turn right. It's six or seven blocks more on the right."

The local fellow is right on. On time to pick up my package, I'm relieved I won't miss a dose of nasal spray that is keeping me symptom-free. I admit, however, to a bit of disappointment there is nothing else in the box—no love-you-miss-you note, no surprise dark chocolate or pumpkin seed treat.

Ah, well, not enough disappointment to quell the thrill ahead. A sign screams from the big blue sky: "Welcome to Montana."

I am encamped beside a lone bush in the flat and barren Beaver Valley Haven RV Park when my phone rings. It's Andy, returning the call I made to him from Beach. "I was out fishing."

"I got my drugs."

"Good news about my aneurysms. They're small, I only have to get them checked once a year. Bad news about Steve. He has stage three lung cancer. They'll do tests in July before talking strategies."

Ugh. I hear the sadness in his voice. Steve is Andy's younger brother. On disability from exposure to Agent Orange in Vietnam, he's had his

share of health problems. Heart attack, cirrhosis of the liver, now this.
"I am so sorry."

FaceTime with Mom, she begs to hear about my day. I tell her about meeting the Newells and my break at Rick's in Sentinel Butte; how Debbie wants so much to help me; Andy's good news, and bad.

I confide. "Sometimes when I'm riding, I think of Dad and start crying."

"I have days like that."

We both choke a little. She wipes her eyes. "We'd better stop talking about it." We sign off with blowing kisses. "I love you."

She posts on Facebook later: "Happiness is having your kids turn out to be nice people."

Funny. Now I want to ride home instead of taking the train. I need to find my own way.

STATS
57.33 miles (1 mile extra to town for groceries), max speed 32, ride time 5:05, avg speed 11.2, TOTAL 1307.28

Me with the Newells at the Painted Canyon.

22

"It's not Montana if it Isn't Windy."

Thursday, June 30, 2016
Wibaux to Circle, Montana

The girl watched her two older teenage sisters. Much of what they did made no sense. Like, their faces weren't broken! Why waste an hour of sleep in the morning to "fix" them? If they didn't like wavy hair, they could cut it instead of using an iron. And what's with that medieval torture device just to curl their already long eyelashes?

It all seemed rather silly. Pick an outfit and get on with it. Why pique Mom's wrath piling clean rejects on the floor?

Nevertheless, the girl stood guard when the two exhaled smoke out their shared bedroom windows. Footsteps on the stairs? "Dad's coming!"

Today starts my fourth week on the road—Michigan, Wisconsin, Minnesota, and North Dakota in three weeks. Montana in two weeks, no problem, yes? Yet I worry. The big sky state is wide as its sky. Hills are such a struggle, and I haven't even hit mountains yet.

Andy would tease me, checking the National Weather Service mobile app on my phone. Every morning at home he consults the "computer" weather on his laptop. I look outside. More often than he'd care to admit, a look outside gives a more accurate version of current conditions.

"It's raining," he might report from the couch.

"Nope, it's clear," I'd counter from the back porch.

Whoot! An east wind predicted today. No sign of any breeze yet, the sun already burns. If the computer weather pans out, instead of a short, fewer-than-thirty-mile day to Glendive, I'm tempted to push eighty to Circle. Save short days when wind fatigue and mountain passes make the ride harder. We'll see. Mornings are slow. Like the Tin Man, my joints need the lubrication of pedaling to loosen up.

Last night's sad thoughts fade. Either I recognize the need to enjoy each moment as it comes, or I underestimate the advantage of a tailwind. It could even be pride in avoiding a disaster when I discovered loose rack bolts during this morning's bike inspection. Whatever the case, today I am invincible. Miles disappear like prairie dogs diving into holes.

As CR 106 dips under I-94 a murmuration of birds swarms like hornets. By reflex I duck beneath their dizzying spirals of chirps and flapping. Above, pottery-like nests hang from the overpass.

"Sorry to disturb you." Pausing there, in the shade and the wildness, I let my mind soar with the flock, free as all the beasts of Eden. No stopping me now. Circle is in my sights.

Glendive before 11:00 a.m., I scarf a steak and cheese sub from Subway (thanks again, Tammy) and veer northwest on State Road (SR) 200. From here, the map profile shows a 1000-foot gain in elevation to over 3000 feet before dropping to 2500 at Circle. Sweet. A downhill boost at day's end.

A cyclist wearing sky-blue Lycra shorts and jersey shimmers from the prairie. White arm and leg warmers cover almost every inch of his skin. A sun-bright yellow handlebar bag and front and rear panniers on his touring bike look like they are about to explode. He slows and moans off a not-quite-broken-in Brooks leather saddle. "Are winds always from the east?"

Ha! Sorry. Not sorry. He's enjoyed more tailwinds than I.

Charles, about my age, is a talkative mirage. Riding to Maine from Vancouver, Washington, today he comes from Circle to Glendive. Of Circle he advises, "The camp there is interesting. It's a trailer court behind a little laundromat."

The uphill pedal to Lindsay is serene, if hot. Sweat drips. No service in this whistle-stop with a population of fifteen, but the ACA update sheet shows a convenience store. This must be it—a decrepit, one-pump-no-overhang gas station surrounded by lumps of scraggly trees. A gust of wind almost wrenches my bike from my hands. Guess the wind powered me, I had no idea it was so strong.

When I enter the building, a squall catches the door, I fight to pull it shut. Oil odors of a working garage waft in greeting. A young man sits on a tall stool behind a grease-stained counter. Half-empty shelves hold dusty bottles of anti-freeze and motor oil cans, a grimy display rack holds a gaggle of windshield wipers. To the right, parts catalogues stained with black thumb prints litter a linoleum kitchen table. A glass cooler offers a welcome icy blast when I reach for an iced tea and a bag of peanut M&Ms from its meager selection.

"Mind if I sit a while?"

The man shrugs. Not inclined to chat, I see.

"Boy, I didn't realize how windy it was 'til I stopped."

This perks him up. "Last Friday we had ninety-five-mile-per-hour winds. Not quite a tornado, so we were lucky. We lost power when some trees came down."

What trees? "Glad I wasn't here for that. Today's the first day I've had a tailwind."

"It's not Montana if it isn't windy."

If the only good thing about headwinds is keeping sweat off your face, I'll take dripping any day. Tailwind and the last ten miles downhill into Circle make for a delightful ride, even at eighty miles for the day. Unable to find Scheer's Trailer Court, where Charles stayed last night, I duck into a convenience store for directions.

"I don't know where it is, but he will." The cashier points to a man behind me. He backs away when I turn. *Do I stink?*

"It's at the end of a dead-end road at the east side of town. Not sure about the owner, though. He likes to drink."

No warm fuzzies. Still, I cycle back to check. There is indeed a tiny laundromat at curb's edge on B Ave, just off SR 200. Deserted, broken down trailers dot the scrubby field behind. A breeze sneaks into the laundromat through the torn screen of a broken window, I can't tell if the clothes strewn about on a table are washed or not. *Creepy.*

I return to town and stop in a hardware. I'll need Coleman fuel soon. The clerk takes me to it, but they only have a one-gallon container. Too much, hoped for a smaller can.

"Do you know where I can camp?"

She drops her head in thought. "No. But, you could ask at the city offices." She walks out with me and points down the main drag. "There, past that pickup truck."

No city offices that I can find, but I remember a Sheriff's office by the store. Surely they can advise me.

"The City Park, it's only a few blocks away," one of two women sitting at desks says. "There's a 'No Camping' sign, but you can camp there. That's just to keep the big rigs out."

The other woman adds, "The pool is right there. You could get a shower."

I must stink.

STATS
83.74 miles (3.74 around town), max speed 33.5, ride time 5:54, avg speed 14.2, TOTAL 1392.08

23

Indecision

Friday, July 1, 2016
Circle to Jordan, Montana

The girl heard the front door close, then laughter. Her parents, coming home after an evening with her aunt and uncle at the Peanut House on 7 Mile Road. She capered downstairs to see them. Her mother must have been in a good mood because she didn't yell, "Go back to bed."

"Patti, what a nice young girl you're growin' up to be," her uncle slurred.

He leaned over her, reeking of cigarettes and beer, and gripped soft flesh above her elbow. The edge of the kitchen counter creased a furrow across the girl's shoulder blades. The kiss came quick. He pressed whiskery thin lips against hers, his tongue swept in hunger. Between her legs a warmth unfurled, at once uncanny and captivating.

And then he was gone. She opened her eyes like surfacing from a dream and heard retching in the bathroom. No one noticed her. In a flash she was back upstairs, slipping under her brother's bed in the room next to hers. She held her breath. He did not come find her.

Abruptly the trickster wind stops, creating an eerie hot quiet in this weird and wonderful landscape. Here, new Murphy's Law rules. Each hill's crest brings a roaring truck or camper. The shoulder is marginal. I smolder, a human charcoal plopped into the monotone terrain of another planet. Wheat grass rolls up and forever, beyond the blue horizon. Can yellow have an odor? Does it smell of cowardice or thirst?

Thirty miles in I reach an oasis. A brand-spanking new rest area with air-conditioned bathrooms. Perched at the intersection of SR 20 (heading north to the Charles M. Russell National Wildlife Refuge), Flowing Wells Rest Area is the only sign of civilization between Circle and Jordan. I take refuge at a cement picnic table under a cement roof under a blazing big-sky sun.

A short, heavyset woman, with more than a few teeth missing, waddles by to the dog pee-pen. She carries a pooper scooper and a black garbage bag. "I hate it when someone pisses in my boss's Cheerios." Reflective sunglasses hide her eyes. *Is she talking to me?* She enters the pen, scoops, and stops on her way back. I guess she was.

"Only three piles. Had to use a gallon of gas to get here. My boss stopped here, then woke me up. 'Get out there right away, it needs cleaning,' he told me. I come out three times a week. I was here yesterday, and it was fine, I knew it wouldn't be much to pick up, it could have waited until tomorrow."

A thin man in a white ACA Transamerica Trail t-shirt runs up and cuts our one-sided conversation short.

"Hello fellow cyclist. Where are you riding to and from?"

My story, redux.

"You rode in '76?" He offers a fist bump. "I started out east to west in '76 but only made it to Carbondale. Finished the rest of it a few years later. Do you recognize me? Michael Prest, featured as a life member in the ACA magazine."

"Sorry, I don't."

A Pennsylvanian, Michael spent the years since 1976 pursuing a goal to ride in all fifty states. "I'm on my way home from the state of Washington, my last one. Cheated the last few states. Drove my truck to get there and just rode around. Rode down the volcano in Hawaii when my wife and I went there for our twenty-fifth wedding anniversary. I rode in Alaska on our thirtieth."

"Why aren't you going to the anniversary?" He mentioned stopping by the ACA headquarters on his way through Missoula. "Mecca," he called it.

"It just didn't work out with my work schedule." He tears up talking about his love for bicycle touring and the original trip in 1976. "I have wanderlust. My dad died when I was born, set my whole outlook on life. You just never know and should live each day on its own merit."

I nod, realizing how nostalgic this ride, my ride, has become. We reminisce about the old days, cycling in cut-off blue jeans and, heavens, no helmets.

"Most of us had toe clips and straps on our pedals," I say, "but I was the only one in my group with actual cycling shoes. I nailed leather strips to the wood soles to make a groove to grip the pedal better."

Funny. Today's high-tech pedals are "clipless." In actuality, cleats on cycling shoes "clip" into pedals like ski bindings, and release with a twist of the foot.

Two hours later Michael says, "I'm sorry to keep you so long, but so happy I took the spur up to Fort Peck. Otherwise I would have missed running into you."

"No worries. Can't you tell I'm eager to get back to the heat and hills?"

Michael's energy revives my battle with the shark-teeth elevation profile of the ACA map. Saltwater sweat sears my scabbed lips. I alternate between rolling and strolling, knowing what I must do to survive.

And then comes a new test—eleven miles of construction. Pavement is gone, traffic routed to a temporary road that isn't a road. It's a two-track, winding up and down through tree-less woods, dodging orange barrels that mark where the dirt is too soft to drive. I am forced to walk much of it. *Am I masochistic at heart?*

Eleven miles drag into eternity, if eternity is time suspended. Grateful for almost non-existent traffic, I keep on with the sense of being watched after. Again. "Thanks Dad!" I yell to the empty sky as I remount on fresh asphalt where the road returns. The contrast in road conditions reminds me of Andy's morphed use of the term "boundary maintenance." Instead of its typical meaning—how societies preserve division between social classes or "others"—Andy says boundaries teach one to appreciate a condition due to experiencing the lack of it, or its opposite. And maintenance refers to recognizing where your boundaries are, and either choosing to not go beyond them, or accepting the consequences when you do. Take fishing, for example. It is one of Andy's favorite things, but come winter, he puts his boat away (and doesn't ice fish). The long break builds his excitement for when the docks get put in in spring.

After those eleven miles, the smooth pavement and slight downgrade into Jordan are glorious.

Jordan, population 343, is a shady town with all services, including camping at the city park. Deep green grass under tall trees, the park is heaven on a side street, less the luxury of an outhouse that I can see. Assuming they are local, I stop a woman and teen walking with a lanky black lab.

"Can I camp here?" I bend to pet the dog. He is happy to oblige.

"We don't know. We're staying in the RV park over there." The woman points across the street. The K & K RV Park is desolate with old trailers like so many other prairie parks. "My daughter and husband and I are here to pick up my son. He's been on a three-week study trip with his high school, on an archaeologic dig."

Archaeology is a thing here?

No one in the home office of the K & K, so I self-register at a weathered wood desk on the porch and slip an envelope with cash through the mail slot. Next to a key attached to a length of heavy chain, a hand-written note informs, "Key to shower and bathroom in cabin across the way."

After setting up near a tree next to the shabby shower house, I take an alley shortcut behind the Garfield County Health Center to a grocery store on Main Street. "Laundromat" is painted on the white brick wall behind the store.

Clean and full, I lug my closet-pannier full of dirty clothes back down the alley. A rabbit bounces toward me. We both freeze less than ten feet apart.

"Bunny?" Nose twitching, he glances up. Slow and sure, he hops a gentle berth around me as if I am nothing but a mirage.

I breathe.

The laundromat is a bust. Smaller than the one in Circle, with no change machines or soap dispensers. Two dryers, loaded but not running, and clothes (dirty? clean?) lie tossed on a table. Dejected, I head back to the K & K. A dog bark makes me jump. A graying black lab, lying in a dog run overgrown with weeds, struggles to stand.

I pause. Another dog, this one brown, sticks his head out from an igloo doghouse. He watches, hesitant.

"Hey buddy, how ya doing?"

Now a cat startles me. Less a rear right foot, she limps over and rolls on her back at my feet. Out of nowhere, three more scruffy cats appear. They circle me. The hair on my neck pimples.

"I think I'll just get going."

I turn and there's Bunny. Facing away, he sits on his white puff of a tail, mottled gray fur blending with the gravel of the hospital parking lot. Evening sun glows pink through his ears. His dark eye, circled in white, keeps watch. He suffers me to approach, close enough that I see his whiskers shiver with every nose twitch.

"Hey there Bunny, my tent is this way, come on over!"

Back at camp, I sit on the picnic table to call Andy. Bunny hops over to hang while we talk.

"I'm feeling the train today, Andy." Not a free-form-go-where-I-feel-like-next adventure, this ride is a niggling pressure. The more pressed to arrive in Missoula on time, the more I think I might be done once I get there.

"I'll look into what it'll take to ship your bike. I can do that easier than you," he says. Ever the supporter.

Meeting the Newells at the Painted Canyon pepped me up to ride home. Today, after talking to Michael about the train, I'm not so sure. "I don't think it's cheating," he said when I brought it up. "A train ride would be fun."

Why should I put credence in strangers' opinions? Mom wants me home. Andy favors a faster return, but insists, "Keep me out of your decision."

Why can't I decide?

I do not notice when Bunny leaves, I only notice he is no longer here.

STATS
68.39 miles, max speed 39, ride time 5:38, avg speed 12.1, TOTAL 1459.52

Me (age three) with our pet rabbit, Sniffy. Did Rick really toss her over the fence to see if she'd land like a cat like I remember? Did she break a leg when it caught on the chainlink? I don't remember when or how she disappeared.

24

Noticing

Saturday, July 2, 2016
Jordan to Sand Springs, Montana

Blue smoke swirled dragon's breath through her mother's nose, spewing argument with the girl's oldest sister. It was horrifying and magnificent.

The slow-motion miasma that obscured Sue's departure from the dinner table hypnotized the girl. Time drawn out with the scrape of the kitchen chair, the dissonant stomp of anger up the stairs, the slam of the shared bedroom door in teenage angst.

At once, the girl vowed there were things she'd never do. Smoke. Fight out loud with her mother. Avoid becoming a mother. Her mother.

Not quite 11:00 p.m., the old-fashioned ring tone of my phone snags me from sleep. "Hey Patti!"

"Mikey!"

Mike Dobies went to college with my brother, Jim. We met in the early 1980s at the Wolverine 200, a twenty-four-hour cycling event on Belle Isle in the middle of the Detroit River. A constant training companion for more than a decade, he crewed our twenty-four-hour women's-tandem ride in 1986, rode PBP with us, and pulled me through many a race. Lou and I call him "Mikey." Like she did me, Lou inspired Mikey to run. He ran his first Boston Marathon in 1993; in 1994, I was proud to be on the sidelines cheering him on. Three more Boston marathons and he kept on running.

"I handle ultra-marathoners now," Mikey tells me as we catch up, having lost touch these last few years. He travels the world to support top runners. "I'm not a good ultra-runner myself, but I know how to crunch numbers."

Not surprised. During my twenty-four-hour stair climb, the now-retired automotive computer geek tracked my pace. I loved when, at some point, he announced, "You've climbed enough stairs to reach the top of Mount Everest."

Before I left Michigan, I emailed Mikey's last-known address letting him know about my adventure. I asked if it was he who anonymously sent me a book after Dad passed. *The Big Dog Diaries: part 1 My Name is Big* by Lazarus Lake is a collection of stories about an abandoned pit bull that adopted "Laz" on his farm in Tennessee. Laz is the brilliance behind the Barkley Marathons, an irreverent ultra-marathon trail race said to be the toughest in the world. Mikey had to have sent it.

"Yes, I sent you that book, but not because of your dad. I put it on my Amazon cart for your birthday, but never sent it. One day I noticed it still there, so I clicked on it."

We stay up way too late talking. I share thoughts about the train. "I don't know what to do about coming home. I change my mind every day!"

"After you are in Missoula three days, you'll know if you're done. And what to do next."

Like offering a wheel when I need it most, sage advice from an old friend.

Our conversation brings up memories that keep sleep at bay. Mom thought Mikey would be a good catch for me, but we never had that kind of spark. Well, at least not me. If Mikey did, he never let on.

In a conversation between us years ago (I don't remember what triggered it), Mikey remarked it was a shame how some people have kids and turn out to not be good parents. "You'd make a great mom," he said, although he knew I never wanted kids.

Motherhood is the most important (and most difficult) life's work. I never felt up to it; never wanted to subject a forming person to my quirks, and truth is, I feared having a daughter who loved dress up and Barbie dolls. What if she asked me how to put on makeup?

Step-kids were enough for me. I concede, though, I could be a better mom now. Witnessing death changes perspective. Things that used to irk me no longer do. Not to say that kids are like puppies, but the scientific-based positive training techniques I'm learning through LBD can apply to any creature. Behavior is just behavior.

Is it wrong to not want kids? Aren't there enough people on this planet? As a thinking human being armed with technological advances allowing choice, can I not enjoy my DNA's drive to replicate and still prevent its consequences? Damn those billboards…

After only five hours of sleep I wake up coughing. My chest is tight. Where is that dang inhaler? I miss the comfort of a humid forest, nowhere to hide from the sun on these god-forsaken prairies. My cracked lips throb. I've tried Blistex and Burt's Bees. I hope this new cold sore cream gives relief.

There are birds here that look like our Midwest mourning doves, but their song is not a soulful coo. I've been trying to give words to their maddening, repetitive screech. Nothing comes to mind. Until I lift my spoon of oatmeal.

"The BROOMstick! The BROOMstick!"

Now I can't get the wailing words out of my head. The BROOMstick! "What?! You think I should take the broomstick home?"

In my notebook I scribble, "Can you tell I am tired? And whiny, evidently!"

There comes a point during every adventure when uneasiness hits. Tired, with no choice but to continue. Wanting, perhaps, to be home. I remember a similar feeling in 1976 in eastern Colorado. Our assistant leader left the group in Pueblo with bad knees and our leader asked me to take his place. "Ack. No." Irritated with the group, I wanted no responsibility. I wondered then, was it allergies? Sometimes allergies get the best of me and affect my mood. Is it allergies now?

SR 200 for at least four more days. No need to look at the map, no need to be annoyed by the profile that shows a gain of over 700 feet. Today I don't care when my handlebar bag sags to obscure my Cateye computer. I ignore it and just ride. A little gray bird races me on a downhill. The road rises, the bird slows in sync. By the time I enlist my granny gear, the bird lands on the white line ahead of me, waiting for the race to begin anew.

"Ramon" fades, overpowered by the BROOMstick. I stop to listen past the doves. Vastness. Foreign. My breathing ceases to rasp, I hear a breeze tickle.

Yet not quiet. Insects voice life stories in songs that course wavelike through the grass. Brown birds chitter and chat and scatter in fanatical flight through heat waves rising from the road. The smell of sweet grass pushes through the dry. Land flows brown and pale green in every direction, broken by a jaded twist where water must rush when it storms. The only sharp contrast across the muted horizon of the range is a dark line of marching cattle.

"Baa," a cry from the distance. A boulder moves. Not a boulder—a dusty gray sheep among a flock of grazing boulders.

Why do you cry so, lamb? It is the same for us all. Live and die. Do what you do in between.

I ride on. In the distance, car tires hum long before I spot traffic. I ride on, unconcerned with lost momentum. *It might take me all day to ride thirty miles.*

The road ahead slices straight to the heavens.

"If there is anything I've learned from Tortoise," Bunny voices in my mind, *"is to slow down. Sense things. See things. You are not done, warrior. You have a lot to learn. It isn't about push, push, push to make a goal. That is artificial. It is about* being.*"*

"I will take the train, Bunny, if I can learn your lessons in time."

An antelope stands in a hollow across the road, caught unawares, antennae-ears twitching, watching, poised to leap. A farmer drives a prairie-sized combine from the other direction as I inch up a yawning, curve-to-the-right climb. The machine is so massive I could sneak underneath it if I dared. It passes, like the opening of a stage curtain.

There, another antelope poses under a cluster of thirsty trees. There, a third antelope stands facing west. He turns as if to consider me, then hops forward, an all-four-feet-curled-up-under-him hopping, hanging between hops like he is on the surface of the moon and not in a ditch in Montana. After 100 feet, he veers left. In one hover, he clears a fence from one dimension to another. His gait changes to a Lipizzan-like trot. Opposite legs curl up by turn. Every step defies gravity. He shifts from trot to hop and back to trot again. I've seen nothing as graceful. At the top of the rise, at the edge of the horizon, he pauses. We share glances, I nod in appreciation. With one hop he is gone.

Movement from the close-side ditch. A tawny fox spooks. I stop pedaling and plant my feet to watch. Zig-zag this sagebrush, zig-zag that sagebrush, for about a quarter mile before he halts. His head revolves to peer at me.

"Good morning!" He startles, dodging to disappear.

How's this for noticing things Bunny?

A bug tickles my forearm. I smack it into oblivion.

"I suppose I should not have done that, yes Bunny?" No response. A short time later another bug alights on my fairing. I let this one catch a ride. "That better, Bunny?"

Good to have more than one chance to get things right. Or, to get things different. Who are we to judge? If a tick gives me Lyme disease, was the tick wrong? It simply does what it does. Does it give me the disease on purpose? I think not.

Thirty miles does not take all day. By early afternoon I reach the Sand Springs Store, a combination one-pump gas station, convenience store, lunch counter, and post office. "Sandy from Sand Springs" is the proprietor, and runs everything. She tells me I am welcome to camp behind the store. "Our restrooms are out front and stay open twenty-four/seven."

I set my tent up under a listing tree between the store and a shed and return to the blessed air conditioning. A young woman and a little girl sit on folding chairs by the front counter chatting with Sandy. Wood shelves line walls of the store, sparse with single rolls of toilet paper, small bottles of dish soap, box matches, bug spray, and other camping supplies and home goods. In the front window a metal shelving unit displays a few tins of Spam, tuna fish, and individual bags of Cheetos, chips, M&Ms, and nuts. A long shelf divides the store front to back. The front holds automotive oil and antifreeze, the back has antiques and homemade crafts. Hand-painted cards by a local artist draws me, I select a pack of six (plus toilet paper and a can of Spam, tomorrow's breakfast—thinking of you, Dad).

"Do you mind if I hang in here a while?" I ask as I pay for my purchases. There is a kitchen table next to a cooler loaded with cold drinks and a sign advertises frozen pizzas and sandwiches. Dinner tonight?

"Not at all."

Sandy sits and chats with me while my pizza bakes. "I worked for the

Post Office for twenty-eight years, only the last two as Postmaster. I'd like to get ten in to get vested."

Sand Springs is really not a town at all, Sand Springs is this store. "There is only one family living near here," Sandy says. The house that sits on the east end of the store's parking lot belongs to a ninety-two-year-old woman, Sandy's partner in the store. "She doesn't live there anymore since her husband passed. She only kept the store open during postal hours but I'm trying to grow it, so I keep it open longer. It's not a money maker, but lots of cyclists appreciate it."

"I do."

"My husband and I have a small farm, 3100 acres with 250 head of cattle."

"That's a small farm?"

"Out here it takes more than ten acres of grazing land to support one steer."

An overhang behind the store gives no relief against the setting sun. It must be ninety degrees at least. I take refuge on a hard metal rocking chair on Sandy's cement stoop, the only shade around. No sweat-lodge tent for me until after dark. An occasional dusty pickup truck pulls up to the gas pump only to speed away. An SUV parks and a woman helps a youngster use the restroom. Otherwise, I have Sand Springs to myself.

Storm clouds gather over barren hills to the south. I reflect on the day's ride. Not been liking this desolate landscape, I miss trees. And shade. And water. But today's observations bring an appreciation of the scale and diversity of life. And our adaptability. *Thanks Bunny.*

STATS
33.2 miles, max speed 32, ride time 2:59, avg speed 11.1, TOTAL 1492.75

25

"Beautiful and Terrible Things Will Happen."

Sunday, July 3, 2016
Sand Springs to Winnett, Montana

In the annual Fourth of July neighborhood parade, the girl wrapped red, white, and blue crepe paper around the tubes of her bicycle frame like patriotic candy canes and clipped playing cards against the spokes for motorbike sound effects. Ending at nearby Fargo Park, the Detroit Parks and Recreation Department put on kids' games. Two fire departments squared off in a water fight, blasting a barrel strung between two poles.

The girl, with short brown hair, lined up for the girls-twelve-and-under, 100-yard dash. A blue and white striped t-shirt wrinkled from the waist of her blue jeans, ironed-on patches curled at the knees. She leaned forward, Converse All-Star clad feet poised for the gun.

"Wait! There's a boy in line!"

The girl looked down the row. No boys. An official ran up to her, the whistle around his neck clicking against his clipboard.

"Me?" She straightened. "But I'm a girl!"

The judge looked her over, shook his head, and waved his hand in a circle to get things moving. A second judge spotted the girl. "Stop! There's a boy on the field!"

Before he could reach her, the girl slammed fists against her hips. Elbows out, she inflated her flat chest and took a defiant step forward. "What do you want me to do? Drop my pants?"

The girl ran so fast she should have raced the boys instead. "I bet I could have beaten all of them, too," she thought as she claimed her prize.

The road from Sand Springs rolls like ocean breakers. I am a short-grass prairie surfer.

"I see you, Bunny, darting from one sage bush to another down there in the holler." I wave, cresting a rise.

"Very good," Bunny says. *"You still need to work on your observational skills."*

Scrubby outcrops reach into the blue like scattered peaks, shoulders gouged in hard folds. Ahead, a descent is obvious. Not so obvious is a distant, pale blue line peeking around the ridge. Mountains? And wait, that darker line, are those trees?

I don't know why I glance left, but I do.

Poised on a wood post high on the brink, a ferruginous hawk, (larger than a bald eagle) peers toward the ground. Compelled to stop, I dig out my Nikon and fire away. A small bird dives out of nowhere and attacks his back. He doesn't flinch. I wade across the road to close the gap, he coils and lifts.

Click, click, click. I capture a downstroke, an upstroke, a middle glide. He soars away with no hurry or concern. "Bye Dad," I whisper to his contrail.

"Nice catch," Bunny whispers back.

Twenty-two miles surfing the troughs and curls of the prairie at an average speed of almost fourteen miles per hour, I arrive at the Mosby rest stop before the road drops to the Musselshell River. Inside, air-conditioning tempers my zest.

A man enters. He folds one arm across his chest and rests the other on it, fist under his chin, to peruse a Montana map on the wall. "There are some natural springs in Great Falls fed by water in the Little Belt Mountains. Here." He points to an area northeast of Helena. "It takes 200 years for the water to get from the limestone in the mountains to the spring."

"That's an amazing piece of knowledge."

"I grew up in Cut Banks and went to college in Missoula. I'm from Bismarck, North Dakota now. Driving to Great Falls to visit family."

Usual questions, usual spiel.

"Are you riding home, too?" he asks.

"Yes." Today's surprise answer. "I'm thinking about taking a southern route back across Montana, along the Yellowstone River."

"That's a beautiful route," he nods, and sweeps his finger across the map. "Have a safe ride."

I plot a course home. Missoula to Helena (I could stay with the Newells) is about 115 miles, an easy two days. Helena to Three Forks, sixty-six. One day. At Three Forks, pick up the ACA Lewis & Clark Trail again to Glendive, 417, making 598 east miles in Montana, forty miles further than the route I am on now. Backtrack the ACA routes in North Dakota and Minnesota, and find my own course across Wisconsin, with a stop in Eau Claire at Karen's (a Leader Dog friend). A straight shot from her place to the ferry in Manitowoc gives a shortcut across the big lake into Michigan. Maybe I could visit Kathie in Ludington, Judy in Clare, and then where? Karen and Ron's in Midland? A Leader Dog friends march across Michigan.

So. How long to get home? Montana in ten days, North Dakota in seven, Minnesota and Wisconsin maybe ten?

Hmmmmm.

Headwinds after the rest stop drop my average to 10.5 miles per hour. No longer surfing, I paddle the twenty-two miles to Winnett. Not much happening in this dusty town of 192 on a Sunday afternoon. At the only grocery store, where the only employee stands outside the door smoking a cigarette, I park my bike against a split-rail fence.

"We close at three." The woman exhales with a puff and snaps the cigarette butt into the dirt. She follows me in to the cool, cave-like interior of the old store. I clomp up and down every dark aisle, half expecting, half wanting stalactites to drip water onto my head. A young woman dashes in to buy a six pack of Big Sky IPA. A beer-bellied older man grabs a case of Bud Lite. Last rush for the Fourth of July I suppose.

The smoker points me to free camping at the city park. "But no bathrooms. It's next to the city pool, open until five and again from seven to nine. Oh, I think the Kozy Kafe is open tomorrow. It's at the corner, across from the Post Office."

The park is easy to find. Compliments of the Winnett Lions Club, it has a swing set and a pavilion with two cement picnic tables. Green grass is a giveaway: "Automatic Sprinklers Used 8 a.m.—10 a.m., Thank You, Town of Winnett." *No, thank you.*

A handful of kids splash in the sparkling water of the city pool. "You can shower for free," the lady lifeguard offers.

"Thanks, I'll be back later. The water looks so inviting I might have a swim too."

I walk two blocks to the Kozy Kafe to check hours for restroom availability. There won't be any hiding behind bushes around here. The Kafe looks like it once was a one-pump gas station. The front roof extends over a cement patio with four redwood picnic tables. Two older men and a woman sit with drinks. The woman leaps up at my hello.

"Do you want to sit in the bar or the restaurant?" She's the waitress.

"What are your hours?" Spontaneously, I add, "I'll have dinner in the restaurant."

"We close at six and open at seven for breakfast," she says over her shoulder, leading me through a central vestibule and the right inner door. The bar is to the left. A half-dozen tables makes the place cozy as advertised. In full view behind a short serving counter, a cook stove stands ready. Except for the men outside, I am the only customer.

"What are you world-famous for?" I steal Andy's line.

"The halibut sandwich is great."

Such a quick answer, I don't hesitate. "That's what I want."

"But we're known for our breakfasts." The waitress sets to work as the cook.

The sandwich tastes fine, but I wouldn't call it great. *Guess I'll check out breakfast tomorrow.*

Belly full, I soothe my aching legs in the pool's crystal water. No more surfing today. I float. Now I notice the mountain range dominating the southwestern sky, with clouds even more mountainous hanging against its peaks. The air glows pink and purple and orange. In the distance, lightning whips sheets of rain in a stampede across the high prairie.

Before sleep I stumble on my brother Jim's sermon on Facebook. *A message about walking by faith and living life. Walk by Faith not Sight, St. John UCC, July 3, 2016, Defiance, OH, 2 Corinthians 4:16-5:10, Matthew 8:22-26.*

Jim speaks of me. Here is an excerpt:

> My parents were reluctant to allow her to go [on B'76], but then acquiesced, finally giving their blessing. Keep in mind that this was long before cell phones and the internet. The only way we could keep track of Patti was through her postcards, and the Super

8mm silent movie film she would send every couple weeks. My parents would take the film to be developed so we could catch a glimpse of her experiences. Today at age sixty, Patti is in the midst of riding alone from her home in Lupton, Michigan...to Missoula on the western edge of the state...My mom is worried about her, but she knows that Patti will live the life she was called to live...

Life is meant to be lived, my friends. To walk by faith means that sometimes we need to stop trying to be in control of everything and trust that the Spirit is at work...Theologian and author Frederick Buechner writes, "Here is the world. Beautiful and terrible things will happen. Don't be afraid."

POSTCARD FROM THE ROAD 7/3/16, about the night previous in Sand Springs

When you are camping, it is sometimes advantageous to have to pee at two o'clock in the morning, even if the bathroom is 170 steps away.

Zippers unzip on sleeping bag and tent door. I stick my feet out into the tent's vestibule and slip them into my waiting running shoes, stretch forward to unzip yet another zipper, then muscle to stand. Emerging from a small backpacking tent should qualify as a tai chi move.

It is dark. Not everywhere I've camped has been this dark, often overhead security lights blare away the night. At first, I see no stars. It must still be cloudy. For some reason, I look straight up. The Milky Way shimmers in all its glory from a skylight opening in the clouds.

Something catches my eye from the west. The darkness is pierced by lightning flashing every few seconds. I have my very own light show before the Fourth of July.

STATS
44.91 miles, max speed 31.5, ride time 4:16, avg speed 10.5 (was almost 14 the first 22 miles! WIND), TOTAL 1527.73

Me, age twelve.

26

Quantum Tricks

Monday, July 4, 2016
Winnett to Grass Range, Montana

Lecture halted mid-sentence, an inhale of breath swept the alphabetically-seated seventh grade IHM classroom. Eyes fierce upon the woman lay teacher (dare rebuke me), the girl pressed her stinging palm against the desktop.

The boy, Brown, sat behind her. His hand had first brushed her arm; she shrugged. His hand squeezed her shoulder; with a razor-thin turn of her head, she whispered daggers. "Stop!" His hand reached her budding breast.

That's when she twirled a tempest across his cheek.

The teacher resumed, and the class exhaled. His hand never breached again.

Phew! What's that smell? If it's me, I must be dead.

Tip for the hot road: don't pack leftover cauliflower. Lucky I have plenty of time to clean up before sprinklers turn on. Unlucky the Kozy Kafe doesn't open for another hour. Another tip: Post Office lobbies, with power plugs, are often open twenty-four/seven. I charge my FitBit and phone while I wait for 7:00 a.m.

I. Really. Have. To. Pee.

To take my mind off my bladder, I trade emails with Karen in Eau Claire. She invites me to stay if I ride home and connects me with her niece, Erika, who lives in East Missoula. "You can stay with her." This keeps my plan to meet Big Dave (and Rich Landers, an independent rider who paralleled our group in '76) at the ACA headquarters in Missoula on the fifteenth.

What about my return? Riding home is doable. Thing is, do I want to? One minute I'm pedaling, the next, a relaxing train ride tempts me, never mind what happened to the free-form, Forrest Gump take-the-long-way-home. Was riding beyond Missoula ever a real option? Perhaps, like trading a no-plan jaunt to Colorado in 1974 for an organized tour with B'76 two years later, my calculated risk this time had always included a straight-away ride home.

My RIDE/TRAIN choices mirror the OPEN/CLOSED sign on the Kozy Cafe door (now flipped to OPEN).

Across the street, Bunny turns tail.

"Mornin', Bunny."

Bunny sniffs. *"That closed thinking of yours reminds me of Turtle, aged wisdom and all that. Be careful, or you'll get tarred up like my long, lost cousin Brer' Rabbit."*

I sigh. Like Brer' Rabbit, I love my home, which we call "the patch."

Last night's waitress was right, this is a happening breakfast place. When I enter, one senior man sits at the table by the door. By the time I order, a three-generation family commandeers a long table and a handsome man wearing a cowboy hat plops a toddler on his knee next to the old man.

The morning menu includes a story about the owner, Ellen, and a 2005 feature on her pancakes in *Gourmet Magazine*. With her dyed hair shaped like a Q-tip (like so many of the ladies that live in Mom's senior apartments), hunched back, and misshapen fingers she must be pushing eighty, yet she takes orders and cooks with no other help.

"You're working too hard," I say.

She pours my coffee and keeps going. "I want everything to come out quick."

Breakfast is quick, and so huge it fills two plates. Cheese omelet stuffed with tomatoes, mushrooms, and green pepper on one, hash browns and toast on another. Even if I wanted, there is no way to finish everything. I drape a napkin over what's left on my plates when a couple, dressed in cycling clothes, nod to me as they enter.

Marcia and Rand are bicycling from their home in Klamath Falls, Oregon to their daughter's home in Newburyport, Massachusetts. "I retired as a family doctor and five days later we were on our bikes," Rand says. "In 2010 we rode from California to Florida to mark my sixtieth birthday."

Marcia rides a short-wheelbase, Rans V-Rex recumbent and is interested in my Tour Easy. We compare notes. My return comes up.

"We made reservations to take a train home. Then found out we'd have to spend twenty hours on a bus through New York. We canceled. We don't know what we'll do, but it'll work out."

Outside, I call Andy.

"I found a place in Missoula that will ship your bike and help you pack it. It's expensive, though, more than $400. If you're taking the train, you'll need a reservation before the cost goes up."

"Seems like a pain to get my bike home. I just can't decide."

"Why not give yourself a deadline?"

Not a bad idea. I peek at the Montana map. "Let's see. I've got eleven days to get to Missoula. Great Falls is about halfway. What if I decide by then?"

"That could work. But remember. Don't include me in your decision. This is *your* thing, I'm your enabler."

"You don't think it's selfish of me to ride home?"

"No. Whatever you do is okay with me. If you ride, though, do you think you can you get home by the time the kids visit? They'll be here the weekend of August 12. They'd love to see you." Andy's youngest daughter, Mandy, and her friend Sharon, are driving from Philadelphia; his youngest son, Josh, and his wife Veronica, will come from Raleigh, North Carolina.

Twenty-six days. "Might be a challenge." Maybe the winds will be with me, maybe I'll be stronger from resting these shorter days and the days off in Missoula. "We'll see."

Riding west from Winnett, I pretend the tears in my eyes are from the wind. A big sky filled with sun, over terrain no different from yesterday, and the day before that, except for the bluing view of mountains marking the horizon, makes for a thoughtful pedal.

What am I doing? I've nothing to prove—my goal to reach Missoula is within reach. Why can't I settle what happens after? Do I want to be done and am afraid to admit it? Or reluctant to concede I may never be done?

Jim's sermon comes to mind, how he spoke of life needing to be lived; his quote of Buechner echoes. I stand fast without fear in this beautiful and terrible world, called to a life on the road. Yet. My life with Andy affirms me.

A mule deer bounds over a fence at the top of a ridge and distracts me. It grazes to the road's edge, scans west, then south. When an eastward glance finds me, ears flicker a freeze. In an instant, he springs up the steep hill to join three others. Before I draw a breath, they disappear.

"Dad, what do you think I should do?" I beg the empty horizon.

No answer.

"If you were still alive, I think you'd say, 'Go home to Andy.' Now I think you'd say, 'Do what you set out to do. Andy will be there when you're done.'"

Grass Range, my stop for the day, is a mile uphill off the ACA route. The greasy, meaty, delectable aroma of a barbecue draws me to the Little Montana Truck Stop Campground. A young man, wearing stained blue jeans with keys on a silver link chain at his side, lifts the lid of a black barrel grill. He ducks Aviator-clad eyes from the smoke. "Ten bucks and you get a hamburger or hot dog, baked beans, potato salad, a drink, and watermelon. Come on over!"

A gray-haired woman in a blue flannel shirt slouches against the building and clutches a change box in her lap. Beside her a young woman sports a red, white, and blue, tie-dyed pullover. The change box holder is the owner of the café next to the truck stop; the girl is her granddaughter.

Greg is the resident cook. "I spent thirteen years in the army with communications and they discharged me on a disability." As I wait for a burger, he launches into a long story about the gray-haired woman backing her car into his, in some other town. "A couple weeks later I came through Grass Range and stopped at the café. It was her! I teased her about hitting my car. She didn't recognize me at first." Eventually she remembered, and they got to talking. No surprise, Greg's a talker.

"She found out I was a short-order cook before the army and offered me a job. So my part-time, on-call position turned into five days a week, in what?" he nods to the woman, "three days?"

She nods back. "But he's good. It was a blessing to find him."

And a blessing to find him today. Hamburger juice and ketchup drips with my first bite. Heaven. Normally not a potato salad fan, this creamy mess turns my heart. I am not inclined to leave.

"You should consider the city park," Greg advises. "It's nice, and it's free. I wouldn't pay to stay here."

"Here" is a half dozen camping spaces more suited for a trailer than a tent, close to long-haul trucks groaning in for gas. Showers in the truck stop and electricity at the sites are attractive, but this morning's cycling couple recommended the city park too. "There's water, although it tasted of iron, and a porta-potty."

I extract myself and opt for the park.

The population of Grass Range is just over 100. Its city park is several blocks from the highway, on the corner of a deserted main street smattered with worn down buildings. Red, white, and blue bunting hangs on a garage-sized pavilion. Tucked inside against two wind-block walls are rusted metal lawn chairs and a faded couch. Outside, a cement picnic table squats near a red water pump. A wooden playscape swing sways in the breeze. Good news: burnt grass means no sprinklers.

On an in-and-out call to Andy, I wander for a better signal. "Wait. There's something in the grass."

"What is it?"

"A bunny. He's stretched out on his belly. Is he trying to cool off?" I edge closer. Not even his ears twitch. "I gotta go, I want a picture."

As I press the button, Bunny hops, and stops in the sun to nibble grass. He doesn't budge when I take a wide berth to the pavilion and drag a chair to write in the shelter's shadow. Minutes later, he's gone. I scooch up, sure he isn't far. There he is, crouched in a narrow band of shade at the base of a light pole.

"I don't know what to make of you, Bunny." Like hawks, these Bunny visits bestow a sense of being watched over. "Are you my spirit guide?"

Another smart phone advantage? Internet access. Dad always said it wasn't necessary to memorize facts. "You just need to learn where to go for answers." I google "bunny symbolism." Like reading a horoscope, website descriptions of rabbit totems cascade into personal coincidence. Rabbits act as guides between life and death—I am grieving my father. Rabbits are emblems of Shamanic journeys—my trip, of course. Sensitive to surrounding cues, intuitive timing helps rabbits grasp opportunities and survive threats. As Bunny reminds me, observing nature is the best way to refine what is already inside me.

"Bunny, I wonder if what I have to learn is something I must teach (or re-teach) myself."

Bunny remains still.

"I don't want to startle you, but I'm going to visit my office," a man says, hiking his thumb at the porta-potty tucked behind the pavilion. I am startled. When he comes out, he asks where he can buy gas. "I'm heading to Hamburg, New York to a yearly BMW convention."

"There's a station on State Highway 87."

He takes a seat on the couch. (If I only I had a nickel...my Lucy "Counselor is in" sign is on again.) Chris is from southern California. "My first wife died of breast cancer at age forty-two. I looked up my high school sweetheart, and we played house together for thirteen years." He describes his house, perched on a cliff with an ocean view. "When she wanted to move closer to her grandkids, I told her we better get married to protect my capital gains. But we never moved. Last month I saw more blue and humpbacked whales than ever before." Next up, he talks about his work. "I lied about my age at nineteen and went to work for the railroad as a numbers clerk, helping build trains in the yard."

I say, "I ran away to the railroad once. Two weeks of conductor training in Atlanta was like a fantasy camp, but three days on the job in the freight yard I knew it wasn't for me. I've worked in men's fields before and tired of having to prove myself." More than that, realities of off-shift work, a requirement to move closer for on-call response, and the eventual grooming as an engineer meant more time away from Andy. Too much sacrifice.

"I hear ya. I'm over seventy now, retired from a constabulary job. I left after thirty years and two days because I didn't like the new boss." Right into his long-term membership in a BMW club. "I've traveled across the country every year for fifty years to these rallies. I'm stopping in Brainerd, Minnesota to pick up my older buddy. We'll go through Duluth, then north of Superior through Canada to Niagara Falls." He learns of my route through Montana on SR 200 and says, "All-in-all, the 200 highways are, by far, the prettiest."

Talk turns to the grizzly in Glacier Park that recently killed a mountain biker. I tell him about the black bear I met in northern Wisconsin.

"Some years back a black bear killed a guy in his tent, northeast of Yellowstone," he says.

"Jeez, that guy's time was up."

He folds his hands behind his head and leans into the couch. "When the good Lord takes me, I'm going to thank him. I've had a wonderful life. And then I'll remind him how much money I've given to his church."

Chris un-kinks his legs and plods to his bike. Silly me, a BMW is a motorcycle, not a car. He dons a black leather jacket and reflective vest, straps on a bright yellow helmet, and wiggles hands into gloves. He waves as he turns the corner out to the truck stop.

Bunny is gone, replaced by mourning doves blasting, "The BROOMstick, the BROOMstick." I want to plug my ears and lie down, but if I go to bed now, I'll wake up at three in the morning. As if on cue, two bunnies bound from a row of bushes, their play a welcome distraction.

Perhaps I am on the warrior's path, as Debbie stated. Perhaps, as Santosh hinted, my journey is shamanic, soul-shifting through the rhythm of "Ramon" and meditative pedaling.

POSTCARD FROM THE ROAD 7/4/16

I'm missing our Fourth of July festivities in Rose City. Usually I walk a Future Leader Dog in the parade with the Rose City Lions Club and the fireworks at dusk in the city park are fabulous! (Thank you, Lions!)

In 1976 our Bikecentennial group was somewhere in Colorado (I think). We camped in the city park and the town included us in its celebrations.

"Big" Dave entertained the crowd with his harmonica; "Little" Dave made his leg disappear behind a ground tarp. I took Super 8 films of Rich demonstrating an Eskimo roll—on his bicycle. And the fireworks display was backlit by a lightning show in the distant mountains... Happy Fourth!

STATS

26 miles, max speed 17, ride time 2:45, avg speed 9.4, TOTAL 1563.79

B'76 riders in Cañon City, CO 1976.

27

Shifting

Tuesday, July 5, 2016
Grass Range to Lewistown, Montana

With the help of her older sister, Sue (and Sue's accordion), the girl taught herself to tune and play a guitar. She loved practicing with her sweet-voiced, sing-along sister.

With two other guitar-playing eighth grade girls, the girl strummed "Kumbaya" and other modern hymns during Saturday evening folk masses at IHM. In appreciation, the nuns gave the girls permission to plan music for a special mass.

The girl wasn't sure what pissed the nuns off more—the electric guitars they fired up or their choice to play the Beatles' "Hey Jude."

The wind, my bane the wind. I cover my ears with my bandana. Like magic, the wind-charm muffles the bluster. *Why didn't I do this sooner?* Ready now for battle under an azure sky, today's challenge includes a 1000-plus-foot elevation gain in twenty miles, and my first pass across the Judith Mountains.

Only thirty-five miles to tonight's destination at Lewistown, I take a leisurely approach to the campaign. Grass plains rise like bread dough into slopes dotted with conifers while mountains blanketed in lodgepole pines color the horizon cerulean. As the road narrows, the horizon fades from view between folds of forest. The pass. Undaunted, I ascend after few spots of trudging; three times I dodge traffic by rolling into the ditch and find sanctuary with trees. Trees!

Riding over mountains in 1976 taught me that though the climb be long and challenging, it ends, with sweet gravity the reward. Ahead the road unfurls. I let the rocket rip.

Heavier traffic and strip malls introduce a town larger than any I've met thus far in Montana. I stop at its Chamber of Commerce.

"Can I help you?" Behind a counter, a woman folds and stuffs t-shirts into plastic bags for a fun run.

"Your sign said 'information.' What info can you give me?"

She drops a shirt, leads me to a wall filled with brochures and picks a thick, glossy magazine. "This has all kinds of things to see and do around Lewistown."

I step back. Don't need that added to my load. "I guess I'm just interested where I can camp tonight." I ask about the fairgrounds, Kiwanis Park, and a mobile home park and campground listed on my ACA map.

"I don't know about the fairgrounds, but the Kiwanis Park is free. It's only a mile west of town."

Tomorrow the route turns north to next stops at Denton, Fort Benton, and Great Falls. Since the mobile home park and fairgrounds are on the way, I head their direction. Bars and shops sprinkle Lewiston proper. Ah, and a grocery store one block from a laundromat.

The mobile park is jammed with RVs and shoddy modular homes and the fairgrounds look deserted. A woman working in the administration building has no clue about camping. I show her the ACA map listing. "Let me check." Bad news. "We haven't allowed camping for quite a while. Can you give me the Adventure Cycling contact info? I'll call and have them take us off."

The free park it is. I backtrack to tackle laundry. While clothes wash, I catch up on correspondence initiated by Mary, a fellow puppy-raiser in Colorado who is moving to Montana this fall. Her husband, David, is a forest ranger stationed in Plains, about eighty miles west of Missoula. David offers to put me up in Plains or connect me with co-workers who live in Missoula. I explain I'll be staying with Big Dave and share my indecision about the train.

David messages, "You're welcome to stay in Plains if you want a couple days to recoup before riding back to Michigan, or up to Whitefish…Give it a thought. I'd enjoy the company. An option to consider is ride to Plains where we could break your bike down to ship. I could drive you up to Whitefish to take the train or Missoula or Spokane if you fly."

Does the universe direct me to adjust my adventure? How cool would it be to hang with a Montana forest ranger before getting a lift to Glacier National Park?

Another message from David: "Mary said you should convince your husband to fly out, then both of you can take the train. I'll be glad to help with your logistics, just call or email."

I chuckle. Andy—who traveled the world in his executive work life and vowed never to fly again? Nope.

Packing confusion aside with clean clothes, I find out what the Chamber lady did not tell me. One mile to the free park is really two miles (uphill), the last mile a pot-holed gravel and sand quagmire of construction. *Should have food shopped before this crusade.*

The grassy Kiwanis Park looks like a freeway rest stop. Toilet building and drinking fountains, no showers, cement tables (two under shelters). Beyond a fence is the Lewistown Municipal Airport. Except for a small RV and an extended cab pickup, a camper hitched to a truck, and two backpacking tents zipped tight near a tree that is more of a bush, no one else is around. Maybe the tents belong to fellow cyclists, off exploring. I pick a sheltered table between them and the bathroom.

A young man exits the camping trailer, waves, and wanders over. Jay and his wife are vacationing from their home in Minnesota. We spend a fair amount of time chatting about my trip and the great train debate. He examines my Tour Easy. "I work for UPS. They would take your bike."

Great.

Even though Jay said his wife was uncomfortable staying here, I have no reservations leaving gear in my tent while I go for groceries. By the time I return, his truck and camper are gone. And the tents do not belong to fellow cyclists—a battered work truck and an SUV park; three men gather at their picnic table with beers in hand. Two attend to laptops, the third works with an electronic gadget.

Am I in for a no-sleep night? Ah, well, dinner.

Two of the men rattle away in the work truck. The third retrieves a black instrument case and a speaker from the SUV, pulls out a fiddle, faces away the evening sky, and bows a single note. He tunes each string. In a pause like a breath drawn and held, the fiddler's bow caresses a long, sweet sigh. A flurry of notes ascends in a frenzy, as if dancing a jig with the wind. The rumble of truck tires on the constructed road fall into time.

I am transported, transfixed. My heart leaps to meet the clouds kissing distant mountains as the song slows a howl at the setting sun. How will I continue to exist in light of the excruciating beauty transmitted from a box of wood?

Clapping hands roust me like a carrot pulled from the earth. I cannot remember where I am. Two older men, a young woman, and a middle-aged couple applaud the fiddler. Where did they come from? I can do nothing but join in. *Ah, Jay, your wife is missing it.*

Jimmy-the-fiddler is a geology undergrad student from Nebraska, studying cliffs with a doctorate student and his advisor. "I was born and raised in Washington, D.C., but my whole family is Virginia." He refers to the Americana music he loves to play. "After camping here a month, I'm ready to go home. My wife and I are expecting our first baby soon."

The older men are traveling to the Canadian Rockies. One expresses a particular appreciation of Jimmy's fiddling. "I play violin and viola in orchestras and small groups at home in North Carolina." Michael wears oversized dark glasses even though it is dusk. "I have macular degeneration. I need a lot of light to see, that's why I like to be outside. I'm taking this trip while I still can. My brother Pat, here, came along to help."

Lanky Pat stands apart. He nods hello.

The young woman is from England. For the past year, she and her husband traveled around the United States in an RV. "We bought it on the east coast." She kicks the dry grass. "We're going back, we have to be home next month."

The couple hail from Houghton, Michigan, in the UP, returning from a family reunion in Missoula. When they hear my story (and dilemma), the man says, "Treat yourself. Take the train."

But.

The chill air stirs my heart, the crimson collapse from day to night arouses my soul, the mournful strains of an Americana melody make comrades out of strangers. What will I miss if I take the train home? It is the ride I love; how can I not want to continue?

STATS
35 miles to Lewistown, 41.31 total with what I did around town, max speed 34.5, ride time 4:17, avg speed 9.6, TOTAL 1605.2

Me and Jimmy-the-fiddler.

28

Only by Moving

Wednesday, July 6, 2016
Lewistown to Denton, Montana

"But I am not sorry!" The girl stamped her feet and faced her mother. "I don't care that I got kicked off. It wasn't a real team anyway. All she cares about is winning."

"You can't just barge in like that. You go right now and apologize."

The girl missed her old coach. He made sure everyone batted and played the field. Sportsmanship, that's what he called it. This new coach didn't even come to the playoff game; she sent a substitute with instructions to "win no matter what." The girl played, but didn't think it fair others did not, even as they outscored the other team.

Mother's razor arm pointed. Not sorry, yet not disobedient, the girl went. The coach held her at bay behind a half-opened screen door.

"I'm sorry, Coach."

"Good. You should be."

The girl raised her chin. "I'm sorry I interrupted your party, but I meant everything I said."

"I think you should write about your beer-drinking geologists," Andy says on our morning call. "I've been thinking about it since last night. You know, how our first impression of people is so often wrong."

"You're right, my reaction was not positive."

"I'll never forget this one engineer I had when I worked at Paragon." Years before we met, Andy was the general manager at the small manufacturing company in Wisconsin. "If you saw him walking down the street, you'd cross to the other side. But he was the nicest guy. If you needed help with something, he was the first one to run out to help you."

"Sort of like that customer of ours who worked for Harley Davidson?" The pierced and tattooed rough of a man was gentle spoken with a huge heart. "And what about Steve?" The not-so-young man dressed in clean jeans and a t-shirt, with kinky black hair not long enough to hide his earrings, came looking for a job at our bike store. "You never wanted to hire anyone with tattoos or piercings. I had to convince you to talk to him." Steve was a great employee; our customers loved him.

I can almost see Andy's face scrunch up as he stutters, "I was wr, wrr, wrong." Always a joke between us, how right he is about things.

Why are we so wary of the "other?" Is it a leftover flight or fight response, a tool of survival when our brains were more instinctual than thoughtful? Or is it something learned?

Our conversation reminds me of a Christmas episode of *Northern Exposure* (my favorite television series). Every Christmas Eve Andy and I watch the award-winning episode, "Seoul Mates." The character Maurice is a conservative, never-married, retired astronaut. In this episode, he mourns his lack of family, yet a surprise visit by a Korean woman, her son, and grandson, upends him. Maurice conceived the son in Korea during the war, but never knew. Suspicious, he cannot accept that his own flesh and blood is not white. In a poignant exchange with the character Chris-in-the-morning (a pseudo-spiritual leader in the fictional Alaskan town), Chris consoles Maurice. He says fear of the "other" is not instinctual, but cultural. Maurice can't understand how this might change his feelings. "It's learned behavior."

We can unlearn learned behavior.

The non-threatening pose of a bicycle tourist strips away societal constructs that inhibit people from opening up to each other. We meet. Neither knows the history of the other, and we find common ground in our humanity. Haven't I learned these lessons on bicycle tours forty years ago and more?

Thoughts roil with my boiling water. *Instead of learning something from Bunny, might I need unlearning?* I lift the pot cover and steam clouds my glasses.

"Would you like to join us for breakfast?" Pat startles me.

I look up through foggy lenses. "Just about to eat some oatmeal."

"Well, eat up and come for coffee. We're driving to the restaurant down the road."

"Okay. Maybe this drizzle will quit by then. I don't mind riding in it, but I hate starting out wet. What's this fellow's name?" A strapping young boxer dances at the end of a leather leash.

"Algonquin Jay Calhoun. I got him from a lawyer friend. Do you remember the old Amos and Andy television show?"

I nod between hot spoonfuls of goo.

"The name is from the lawyer character on the show."

I guess I don't remember it that well.

A livelier-than-last-night Pat grills the waitress, "I have a question and what you tell me will decide if I eat here."

The waitress shifts her weight.

"Do you have grits?"

"Nope."

"What am I supposed to do now?" He teases, but orders.

Michael shines his iPhone flashlight to read the menu. I'm glad Pat drove, and I hope Michael takes a different route than me when they leave Lewistown. People often ask if I am nervous of traffic. "Not trucks. Truckers know where their rigs are. It's the RV drivers that worry me."

"So, what is a puppy counselor?" Pat saw my Leader Dogs for the Blind card when Michael and I exchanged contact information at the fiddle concert. "In case you need help on the road," Michael said.

"I help other volunteer puppy raisers. Right now, my group is a bunch of inmates in prison."

Pat asks questions and nods when I describe the win-win-win. "Sounds like a wonderful program. I worked ministry in a prison for fifteen years."

The waitress brings the bill and a raffle ticket for the café's monthly drawing. "You get one ticket for every $10 spent. Put your name and phone number on the back."

Pat looks at the bill. "It's $19.75. You mean I don't get a second ticket?"

"You need to spend a quarter more."

He slaps the table. "Okay, then I'll order a cinnamon roll for Algonquin. But what about last night? We had dinner here and spent more than $20. Got no tickets."

"I'll talk to the owner." She returns with the roll and three more tickets.

"By the way, what's the prize?"

"An AK-47 rifle," she answers, as if an automatic assault rifle is not an unusual prize from a family restaurant.

"If I win it, can I do anything I want with it?"

"Sure."

"Like destroy it, or melt it down?"

Without hesitation the waitress replies, "If you don't want the rifle you can choose something of equal value from the store providing it."

Pat leans back with a wily grin. "Oh no, I would definitely take it. And definitely destroy it."

The sky breaks, my plan to dawdle worked. I wave goodbye to the gentlemen brothers, glad the muddy construction to the ACA route is downhill.

The terrain north rolls long hills into stark vistas ringed with mountains. I classify climbs by how many steps I take to push my bike to the top. This one is a 400-step hill, the one before that was a 600-stepper. Every 100 steps I pause to savor big sky views. And gasp a modest rest.

With the Judith Mountains behind, the road snakes easy amid a broad prairie between the North and South Moccasin Mountains. Do I detect a slight tailwind? A quick peek at the map profile inspires a harder push—a gradual loss of altitude. For a while. Then, a wall to scale onto the plateau that houses Denton, tonight's stop.

I first met plateaus in 1976, when riding a bicycle across the country taught me more about geography than any schoolbook. This plateau peaks sharp before a 300-foot drop to the tableland. An ominous sky and dropping temperatures prompt me to layer up with tights and rain jacket. By the time the road levels out, rain spits.

In a land devoid of inhabitants, the long dirt driveway of a farmhouse with outbuildings beckons me to safety. My quick turn spooks two horses in a wire-fenced paddock. Is that thunder, or their hooves as they race behind a weathered barn?

Hammering echoes from a smaller barn next to my refuge under the eaves of a two-car garage. The noise stops. Dressed in overalls and a John Deere baseball cap, a farmer walks my way with an Australian sheep dog at his heels.

I wave. "Hope you don't mind me taking shelter for a bit."

"No, go right ahead." He ducks into the garage.

The Aussie trots over and begs with auburn eyes. "Hey buddy." I bend for a scratch. He drools for my granola bar. "Sorry, it's mine."

The farmer, armed with a bigger hammer, asks about my ride, even as he edges away. I give him a short version and add, "Heading to Denton today."

"Oh, that's only about five miles," he says over his shoulder. "We need the rain. We needed it in April. Didn't have any then, it wreaked havoc with the hay."

The Aussie hangs near. Rain peters, blue sky pushes clouds apart, the horses venture a few steps into the open. A dusty gray cat trots over and plops under my rear pannier, tail twitching as if I've known her all her life.

"Mind if I take your photo?"

She turns her head in a regal Egyptian pose.

"So long horses," I nod, dodging mud puddles. No spooking this time. As I turn right out of the drive and look up, a dark western sky spooks me.

"Five miles he said? The map shows at least seven. Five. Seven. Ah, well. Can't take too long."

I pedal on.

At the crest of the next hill the sky morphs into a colossal black wall stretching from one corner of the desolate horizon to the other. Stephen King could not imagine a more formidable, breathtaking vision. I half consider stopping to dig out my Nikon, but even my 18mm lens could not capture the entire spectacle. Perhaps a panorama with my cell phone?

Nope, the wind slaps my face freezing. My mind whirls to a day in 1976, crossing Kansas.

A storm chased up behind tandem riders, Len and Loree, and me. Like an upside-down ocean, the sky roiled countless funnels only to retreat into its surf and spooked the native Californians. I was rapt.

"As long as the wind doesn't turn cold, we should be okay." Before the words were out of my mouth, the next blast of air was like someone switched the air conditioning too high. The air turned green. Not even a ditch for cover, the tallest contour for as far across the prairie as we could see was a grain silo marking the next town. We dug deep and rode hard. The tailwind assist carried us to a café just in time, we never even got wet.

Like the violent summer thunderstorms I'm familiar with in Michigan, this one blew through before we swallowed the last bites of lunch. No twisters touched ground, as far as we could tell when we got back on the road. The air above the bowed wheat fields fairly sparkled.

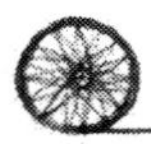

Today, I head right into the clash.

I consider turning around. Can I outrun the tempest back to the farm? Might be closer to aim for Denton. The road bears straight toward a lightening patch in the darkness. Perchance a breach?

I hammer down like Frodo marching into the depths of Mordor. With no warning, a freight train of wind slams into my fairing, spewing rain so vicious my face feels assaulted by the quills of a radioactive porcupine. A semi-truck barrels from behind, water sprays from all sides. *Get off the road. Now.* I roll down a slight slope and tuck as best I can beneath my fairing. I can't see a thing. The cold is dead-of-winter cold. I could wrangle my ground tarp from under my gear, but fear if I get off my seat, the wind will toss my bike to fly with the wicked witch. I hunker tighter. Who would think a piece of Lexan could prove so useful?

A rain this hard shouldn't last long, but here, who knows? How long do I cower? Twenty minutes? Two hours? Shivering, I squint through the fairing. *Is the sky growing lighter?* I poke my head up for a better view. The rain still stings but the porcupine is no longer radioactive, and my wrap-around sunglasses give some protection.

I must move.

Head down—pedal, pedal, pedal. Luckily no traffic passes. I think I'm in the middle of the lane. The rain weakens. A mile from Denton a curtain lifts and the sun shines.

Last night I wanted to ride home, this morning the train. I rationalized Missoula as an appropriate "ending," the train another adventure, a bonus for Andy and Mom. But the instant I realized that only by moving would I survive the onslaught, I decided.

"Okay Bunny. I am not done."

I am riding home.

Denton presents bigger than a town of only 255. A groomed ball field has a grandstand, a grain elevator looms over a rail yard, its main drag sports a library, bar, church, restaurant, and an imposing brick-sided bank with an electronic sign that flickers the temperature at fifty-five degrees.

Something draws me to the library. I dig out a dry top and fleece and enter, shoes squeaking puddles with every step. A woman with a bob of graying hair sits behind a low desk just inside the glass doors.

"Can I help you?" When I ask for a restroom she points me down a narrow hall. "Take a left, second door on the left."

Warmer, I inquire about WiFi. The woman nods yes. "Care if I sit awhile?"

Even though I still drip water from my tights and cycling shoes, Sandra the librarian offers me a seat anywhere. I choose the spot least likely to be water-damaged—an office chair on a tile floor by a row of computers. As is typical, we talk about my journey. My counselor sign must be lit, too; Sandra shares her life story. She is forty-eight with two boys (eight and eleven) who are visiting their dad in Arizona.

"We divorced a year ago. We used to drive Airstreams around the country to deliver them. I loved meeting people. My husband criticized me for that. I was a high school teacher, but now I'm going back for my master's in counseling."

Synchronicity again. "You remind me of my younger sister. She was a teacher, too. She took her three girls out of an abusive marriage and got a master's degree in counseling."

For two hours we share bits of our lives. Sandra has a rare and progressive lung disease and uses oxygen. I have asthma. Her librarian job is new, she has big plans to make Denton's Public Library a gathering place. I am so lucky to have a husband who understands my need to ride.

"I always wanted to be a mother," she says.

"Not me, being an evil step-mother is enough."

"You remind me of one of my friends, she's an accomplished runner and adventurer. She never had kids either and married an older man retired from the army." This friend's husband isn't keen to go with his wife on her adventures, so she does them alone.

"My husband is twelve years older than me, and he's retired from the army too."

"That's it! You were meant to come in here."

My evening call with Andy warms me (forty-seven degrees predicted for tonight). "What are you thinking about your return given you got caught in that storm?"

"I'm riding home. Tired of changing my mind and stressing over the decision. Riding, I can put to rest the myth about prevailing west winds. And besides, it will be all downhill. I've been climbing since Fargo." Maybe to soften the blow I add, "After three days in Missoula, if I find I am done after all, I'll call the ranger. Even if I spend a week in Plains before I can get a train ticket, well, that's what I'll do."

But. I'm riding.

STATS
41. 46 miles, max speed 34.5, ride time 3:47, avg speed 10.9, TOTAL 1646.60 A 1600-step day.

Entering Kansas, 1976. Left to right, eleven-year-old Colleen, me (group cooking bucket strapped to my gear), tandem rider Loree, group leader Steve.

29

Stark Reminders

Thursday, July 7, 2016
Denton to Fort Benton, Montana

The girl went into training, eleven city blocks a mile. The neighborhood kids teased her as she pedaled laps back and forth. "You're crazy, you'll never make it that far!"

"That far" was ten miles one way to visit her aunt and grandmother at their condo in Farmington. It was the summer after eighth grade graduation from IHM and she talked a friend into joining her.

The morning of their great adventure, Sharon rode one mile to the girl's house; together they continued on side streets. Aunt Mary made lunch for the girls and treated them to an hour of horseback riding at the stable across the road.

Sore butt or not, the girl kept mileage equal by riding Sharon home. Her last mile, legs turned like rusty bolts with stripped threads. Turning onto her street, she pumped her heart out to prove the doubters wrong.

All night a generator rumbles from a wide-load semi parked across Broadway Ave. Guess the driver is keeping warm. After an early jaunt to the toilet, I am glad to tunnel back into my sleeping bag. As if on cue from yesterday's musings, the Internet is ablaze with police shootings in Texas.

"Let's not talk about this," Andy says when I bring it up during our call. "My blood pressure is going up."

My heart swells. This man, this gentle man who aims for the win-win in every situation; who assured me when I worried about his kids surviving their teen years, "Statistically, they will be fine;" who hugged me tight as I collapsed in his arms after weekly care visits with my parents. This man thinks gun mentality, or value, or whatever it is, simmers under the surface of our culture. "Like that family restaurant raffling off an AK-47. Who needs a gun like that?" He sighs.

When I get moving, I am glad I cleaned the bike and lubed its chain last night. Numb fingers thaw around a tin cup of hot coffee, steam defrosts my nose while pancakes sizzle. The tattooed truck driver strolls down the sidewalk to the restroom. On his way back he stops to ask about my trip. He is very nice. I secretly forgive him for running his generator all night.

The road out of Denton follows a three-mile arc along tracks where a resting train blocks the view of the distant Highwood Mountains. Before the curve ends, the wide-load caravan fills my mirror. I pull off on a dirt drive that blends into the prairie, the driver salutes with a series of air horn blasts.

No longer cool, the sun beats on my skin like an open palm on a tight bongo, keeping time with the ever-present "Ramon." For two miles (at forty miles per hour), a delirious descent off the plateau through Arrow Creek Hills empties into a depression, another world of rock bluffs and nothingness. *Wait. What is this forgotten phenomenon?* A row of trees cast shadows through a short curve—shade, glorious, cooling, caressing shade. Haven't ridden in shade since Minnesota.

The pleasure is short-lived.

Making fair time, thirty-five miles before noon, another twenty-seven to go. The map profile shows a gradual climb for fifteen miles with a delightful 600-foot drop into Fort Benton, today's destination. What drivers there are wave with two fingers and a thumb, their hand never leaving the steering wheel. I take to waving back the same way, lifting just my fingers from the handlebar. *Remember barefoot drivers in Minnesota?* Bunny probably won't like me imagining all Montanans in cowboy boots.

I crack a smile. Literally. But it's not my wrecked lips that wipe it away. A stark, white cross juts from the edge of the road. Another along a fence line, three in a row at the next bend, two more at the crest of this hill, five at the bottom. Must be a dangerous stretch. An apt reminder for drivers to pay attention, the Montana Highway Department puts markers up where people die in crashes.

Bunny whispers, *"Yes, pay attention. Pay attention to what is important, before it's too late. Pay attention to your loved ones before they are gone."*

"Don't need the reminder, Bunny." Last summer, driving south on I-75 for my weekly visits to care for Dad, a billboard near the Zilwaukee Bridge reminded me too often. I hated that sign, blazoned with year-to-date traffic deaths on Michigan highways. One week the total was 349, the next week 366, the week after that, 398. Dad was dying and I couldn't even say the word.

I pedal on, despite a surge of sadness.

Fort Benton can't be far, but nothing rises from the golden fields of wheat to signal a town of almost 1500, nothing but a long, lonely line of telephone poles stretching west. The road takes a turn and plunges. A gorge splits the prairie; sunlight glistens off a river below. Fort Benton, poised on the banks of the Missouri River, nestles between high plateaus, protected from raging winds.

Immediate attraction.

I cross the Missouri over a bridge upstream from a weathered railroad trestle and cruise into the town's sleepy historic district. Along the river, a levee bears a shady walking path. Less than a half mile further, between Old Fort Benton and the city pool, is a city park with free camping.

Lofty trees shade a flat expanse of green grass sprinkled with wood picnic tables and a few gazebos. On the outside of a brick toilet building, a map shows sprinkler zones and schedules. I can't orient the location of these zones (and where to not pitch my tent). I seek the help of a woman playing with a gaggle of toddlers. Happy to oblige, she can't figure it out either.

"Wait." She points to faded white letters stenciled on the brick above the sign. "The sprinklers don't run on Friday through Sunday. Today is Thursday, so you should be fine tomorrow morning."

"I could hug you."

On my way to the grocery store I spot a familiar RV parked in front of Old Fort Benton. Sitting on a park bench is none other than Michael. He doesn't recognize me through his wrap-around sunglasses.

"Hey Michael, it's patti."

"Oh, hi Patti! Pat is over there getting ice cream. Do you want some?"

Why are these two always trying to feed me? “No thanks, I’m heading for groceries.”

The tall brother strolls across the street with Algonquin’s leash wrapped around one wrist and an ice cream cone in each hand.

“Patti! What a surprise.”

“Hey there, Algonquin.” I scratch when he lopes over for a sniff.

“We plan to spend tomorrow on a float down the Missouri,” Pat says. “Why don’t you join us?”

I am tempted, Fort Benton needs exploring.

“Aw, thanks, but I need to keep riding. Great Falls for me tomorrow.”

STATS

63.79 miles, max speed 40, ride time 5:31, avg speed 11.5, TOTAL 1710.41 400-step day.

30

Floating Into Battle

Friday, July 8, 2016
Fort Benton to Great Falls, Montana

All eyes were on her. Pat got up from her seat in the back of the tenth-grade geometry class and inched her way to the chalkboard. She knew how to solve the problem, and she recognized at least three routes to the answer. But when she turned around, her voice went AWOL. Sweat beaded, she could not swallow.

Silence.

Except for the light tapping of a pencil eraser from the teacher who sat at the girl's desk, the teacher who once told her, "If you dressed like a lady instead of wearing jeans, you'd do better in school." She got all A's. How much better could she do? "If I wore a dress, I'd worry about how I was sitting and never learn anything."

And now she couldn't speak. Why was she so nervous, what did she care what anyone thought? She whacked her palm on the teacher's desk. "This is stupid!" Everyone, including the teacher, flinched. Pat twirled, scratched across the board, writing each theorem as fast as she explained it. She took the longest, most convoluted path to the answer and slammed the chalk onto the ledge.

The map profile of the route to Great Falls from Fort Benton shows a climb to over 3500 feet, followed by a series of serious spikes. No stove this morning, I down a banana and protein bar. I'll grab a yogurt and Gatorade on my way out of town.

As I wolf yogurt in front of the grocery store, a couple walking a young Vizsla along the river path waves. "Nice pup!" The man yells back, "Nice bike!" They cross over to chat. Donna and her husband, Scott, and Zoe the pup, are from Kansas, here to visit their son and his family.

"He's a park ranger," Donna says. "He loves to mountain bike." As if on cue, the son rolls up on his bike. She introduces us. "Patti is riding to Missoula from Michigan. She stayed at the park last night, I wish we ran into her yesterday."

Donna must feel the same instantaneous friendship. Something about this town. Ah, well, the road calls. I steel myself for the climb out of the Missouri River watershed. No oomph in my legs today. *Can I keep track of pushing steps?*

I do the best I can.

In the fewer-than-200-person town of Highwood, a downhill eases past a "Welcome Bikers" banner across Elmo's Highwood Bar. Before the road rises, I turn around. A delightful rush of cool air greets me at the door. Blinded away from the blazing sun, I ask a shadowy form behind the bar, "You open for lunch?"

"Sure, come on in!" A young, blond woman materializes.

"I saw your welcome sign and had to stop." I take a seat at a rustic pine table against one wall. "Riding to Great Falls today. Isn't that about thirty-five miles from here?"

"Not sure. I can tell you about the route though. You've got a pull to get out of the bench, and then you've got a couple of coulees. It's kind of up and down."

Is this another country? I don't understand her words but I have the gist of it. Climbs. I settle in to peruse the menu and charge my phone.

"What are you world famous for?" Again, I steal Andy's line.

"Definitely our hot slaw."

"Okay then, I'll have that and a BLT. If you can burn the bacon."

The house specialty is purple cabbage laced with bacon, Feta cheese, and homemade poppy seed dressing. Even if I wasn't a ravenous cyclist, the dish is amazing. (And the bacon is perfect.)

Two hours later, back to the blast furnace.

What the bartender didn't warn me about was road construction out of Highwood. No pavement, my cycling shoes slip on loose gravel for 500 steps. At the top I breathe and remount. Pavement ahead looks lovely but disappears again on what would have been a nice descent to Belt Creek. I pump my brakes to keep control. Across the creek, back to counting steps.

Eventually pavement returns. Riding on, I scan the wide-open plains for the right turn that leads to Great Falls. In the distance, a road leaps from behind a barren hill and up a barren wall like a metal measuring tape extended to the sky. *Is that the road?* I don't want to believe it.

Sure enough, at the turn the road goes straight up. I am tired of walking. "Ride up it you fool!"

I bully the granny gear at five miles per hour for what seems like hours, the wind a hot slap in my face. Grind on. Ridges on both sides of the road enlarge and close in. Breath comes hard.

Movement catches my eye—what looks like an over-sized dandelion seed floats smack dab onto my sweaty brow. A burst of air launches tens of hundreds of the umbrella-like hairs, the sun illuminating each one into a sparkling white sear. Those cannot be seeds, they are thousands of miniature paratroopers wafting into battle, more and more coming and coming.

Light and bravery lift pedal stroke after pedal stroke. Pedaling gets easier, I crest and hoot and holler into emptiness. "I made it! I made it!" Two deer freeze a stare on the opposite slope. I can't help myself, I pump my fist. "I did it! I rode all the way up!"

Screech! I startle, my celebration interrupted. Another screech. There, perched upon a wooden post like a chainsaw carving is a ferruginous hawk, its dark-streaked eye peering into my heart.

"Hi Dad!" Tears well. Does it matter if I imagine this wild bird is him watching over me? His energy is within me and all around me, as mine will be everywhere one day too.

And here, where acres of grain stretch in waves past an earthly horizon. My mind can't take the eternities my eyes see. Sometimes blue-green mountains peek above the gold, other times nothing but the cloudy sky demarcates the edge of the world. Power poles lining the highway lean this way and that, cross-arms catawampus straight out of a Dr. Seuss book.

Like Dorothy and her band of travelers crossing the poppy fields to Oz, paradise lures me. I disembark to wander into the land, my hands brush silky tassels. If I should lie down, I will never get up. I stay standing. Eyes closed, I listen. Dry grain, whisking against itself in the wind, transforms into rain, the rain in the hardwood forests of home.

Near Great Falls, with a population pushing 60,000, traffic wrenches me from reverie. The River's Edge Trail, a paved path along and over the Missouri River, gives relief past green picnic spots and cement skate parks. My destination is off route on the west side. Unexpectedly, the path, now a wide sidewalk, delivers me right to Dick's RV Park.

Tucked behind a maintenance barn and a dumpster, a split-rail fence segregates a suburban-lot sized square designated for tents. Two cyclists sit at a picnic table. Their bikes lean against a redwood privacy fence behind their tent.

Bedraggled, I drag my bike in a daze after a 2200-step day (I kept track!), unable to decide what site to claim. The couple smiles an understanding look. To delay a decision, I greet them. They are from Germany, on their third long tour, this one from Alaska to Argentina. I sigh, a bit deflated. The woman shakes her head, "No. You have great tour."

Her kindness comforts me.

The couple points me through a back neighborhood to a too-many-choices grocery. I return with my best dinner yet—blackened salmon and a crisp salad.

And Bunny pays a visit.

"Well, Bunny, am I un-learning anything? Or maybe I'm coming to accept the life I've become, that it's okay for dreams to change…I'm not taking the train, you know, so I'll look for guidance riding home."

His nose never stops twitching, eyes never blink, ears alert and never sinking. A long bit later, he circles the Germans' tent and disappears.

STATS
59.11 miles, max speed 40, ride time 6:02, avg speed 9.8, TOTAL 1769.54 2200-step day.

The edge of the world.

31

Busted

Saturday, July 9, 2016
Great Falls to Simms, Montana

Her mother was pissed. She had just scheduled hysterectomy surgery when the father came home with the news. She knew an unscrupulous boss stressed him, that running the factory in this little town wasn't working out the way he hoped. "But right now?"

"I couldn't take it any longer. I'll get another job."

The girl admired him, even though, while he commuted to his new job in the city, she spent her tenth grade summer tending to her mother and three younger siblings. "Takes balls to admit enough is enough," she thought.

Bunny is back. He doesn't twitch when I almost trip over him coming from the bathroom. A FaceTime request from Mom interrupts, her eyes puffy from tears. "Aunt Ag died this morning."

"Oh Mom, I am so sorry." Aggie is Mom's oldest sister. She was ninety-four and lived in a nursing home; her husband, Bob, died some years ago. "What happened?"

"Cheryl called me yesterday. Ag fell out of her wheelchair and broke her arm. I called early this morning to see how she was doing. Cheryl answered the phone crying. 'Mom just died and I can't talk right now.' She hadn't even called her brothers when I called. I prayed last night for her to go, she couldn't take pain pills, and she was suffering so much."

I listen. There is nothing I can do.

"I worried about you last night when you got in so late. So, don't mind me bugging you."

"I love talking to you every day, Mom. I hope we keep it up after I get home."

"When are you getting to Missoula?" Her voice cracks.

"I need to be there next Friday, that gives me six days. I'll get there okay."

"Well, I'll let you get going. Be safe!" We blow our usual kisses, love you, love you too.

Another loss. My heart burns for Mom, who reads the online obituaries from the Marquette Journal every day to find out who died. She tells me I'll know what it's like to live to be so old, looking after her and Dad. "I never knew what to expect," she said. Her own mother died young of a stroke, and while her father lived into his eighties, his second wife cared for him. Mom rarely saw him.

I need a few moments. Should I feel guilty I'm here and not with Mom? *Where's Bunny?*

Out of Great Falls, a mostly level interstate access road treats me gently, with only a moderate headwind. A rider cruising east on a Tour Easy gives a hearty wave. I lift my fingers in reply. Not too many miles later, I imagine the eastbound rider enjoys what I now fight—a fierce headwind. I pass through Fort Shaw so slowly I hear a first-grade nun's voice in memory, "You're slower than molasses in January." Scenes of a park wedding float by: white plastic folding chairs sprout like tombstones and children chase between them; folks dressed in Sunday best gather in clumps; a bride's veil swoops and swallows her new husband's head as they pose for a photo.

SR 200 leads me six more miles into the shade of tall trees marking my day's destination: Simms. What's going on in this town of 354 people? Cars and pickup trucks are parked everywhere with no one in sight. I stop to check the address for the Curtiss Service Center and Camping. It should be just ahead. As I clip my foot in the pedal to continue, a man leaps out of a car parked across the street.

"Hey wait! I was taking a nap and just woke up and saw you. Where're you riding to? Where d'you start? Are you heading to Missoula for the anniversary?"

Missoula…Michigan…yes.

"I was mountain biking about four hours west of here and stopped to rest before driving the rest of the way home. I'm going to the reunion too. Driving some friends to Missoula on Friday. They're starting a tour west from there."

John and his wife are Warmshower hosts in Great Falls. Cyclists I've met have told me about this free, online hospitality service for touring cyclists. Members register as either a host or guest and connect through the www.warmshowers.org website. Maybe because I haven't grown up with the Internet, or prefer face-to-face encounters, I haven't used the site.

The Service Center is at the west edge of Simms. Inside I find a balding man wearing clean blue jeans held up by a wide leather belt and silver buckle the size of a small plate, and cowboy boots so polished I wish I kept my sunglasses on, sweeps the floor. A similarly aged woman in a gingham dress is behind the counter.

"You can camp on the grass behind the station," Gingham Dress says. "But we close at 4:00 and there won't be any bathrooms for you. And we're closed all day tomorrow." Tomorrow being Sunday.

Shiny Boots leans against his broom. "You could check out the city park in town, but not sure there's a restroom there. Or, a campground near Lowry Bridge is about five miles west. There are bear boxes there. Don't mean to scare you, but three weeks ago we had two grizzlies come through town. Somebody saw one head back to the hills right away. The other one is probably gone too, by now."

Not meant to scare me. Sure.

"Or, there's a cowboy reenactment happening this weekend between there and Augusta. I'm sure they'd love to have you stay with them," he adds.

Might as well tough it out to Augusta. Clouds darken the western sky and, last I checked, the NOAA weather app showed a storm brewing.

"The city park sounds fine. I'm done fighting the wind today."

Shiny Boots returns to sweeping and I wander the barren store in search of dinner. A hulk of a man, dressed in a black checkered shirt, black jeans, and a white cowboy hat that looks too big for his head, enters. He grabs a case of beer from the coolers.

"Going to the funeral?" he asks Gingham Dress as she rings his beer.

"Soon as we close."

I listen in. Seems the entire county came to town for a local man's funeral. He was the same age as Aunt Aggie. *How's that for synchronicity?*

Outside, Cowboy Hat lumbers into a jacked up, four-wheel-drive, extended cab Ford pickup parked inches from my bike. He fires up the engine and lets it idle. I doubt he sees me. The engine rattles away with no sign of shifting into drive.

Come on, pop one open, take a swig, and go. I guess I'm in the wild west now because that's exactly what he does.

The park is cottonwoods and green grass, playground, and a porta-potty. I lean my bike against a chain-link fence surrounding tennis courts, near a thick bush offering some shelter from the wind. I unroll my tent and stake the four corners.

A sound alerts me to water. *Damn.* Sprinklers, and one right under the tent.

Bike already out of range, I time my run to grab food pouch, sleeping bag, tent, and tarp in one swoop. I forgot the reason for such green grass. Mounds of dirt criss-cross like mole super-highways—a new installed sprinkler system. Great. No telling the timing schedule, now where do I go? Across the way is a high school football field, but there's no protection from the steam-rolling wind chasing a storm this way. The sky reminds me of the *Game of Thrones* wall outside of Denton.

At the corner, a wiry old guy on a four-wheeler holds a beer and chats with a young man in a pickup truck. An empty gun rack fills its rear window. I overhear talk about the funeral.

"Hello! Do you think anyone would mind if I set my tent up over there?" I point to brown grass next to the school and explain my sprinkler predicament.

"Honey," the man says. "Set up anywhere, no one will care. Heck, you can set up in my backyard and use my bathroom if you want."

I decline but visit a while as they share stories about the deceased. The young man says, "I worked with him on a job one time and complained how hard it was. He said, 'Want to talk about tough? This ain't nothin'. The depression was tough.' What could I say to that?" He shakes his head and lifts a beer in a toast.

False starts finding a spot where the surface isn't like cement, I manage enough stakes in the ground to set up my tent between two wings of the high school. Dinner is a tuna wrap, Cheez-its, and a sweet tea from the Conoco.

I slip into my tent to wait out the coming storm and try, once again, to read *Angle of Repose* by Wallace Stegner. An Adventure Cycling blog post recommended the Pulitzer Prize-winning novel as a "must read" for travelers of the west, but the 632-page brick almost got donated to the Honey Hub. Now the story draws me in. I don't even notice the light dying.

What I do notice is car tires crunching the gravel drive. A door opens followed by a man's voice, "Hello!"

I crawl out. "Guess I'm busted."

Under a visor cap and dressed in a gray t-shirt, khaki shorts, and sandals, the man reaches out his hand. "I'm Dave, school superintendent."

"Did someone call you? I tried the park but the…"

Dave cuts me off. "I just stopped by to check on the new sprinklers. Come with me."

Sure. How likely is it for the superintendent to check sprinklers on a storm-threatening Saturday evening? He unlocks a metal door and sticks a small stone in the jamb. "If the storm gets crazy tonight you can come in here. Just remember to kick the stone out when you leave. Come on in, I'll show you around."

He listens to my story.

"Right on! Too bad I didn't know you were coming through. I live in Fort Shaw, one town back. The other side of my duplex is empty. You could've stayed there."

"This is so nice of you. I can't believe you're giving me the school tonight."

"Right on! Happy to."

We cross the gym and Dave leads me to the basement locker room. "You can take a shower here." When he can't unlock a gated area, he apologizes for not getting me a towel.

"That's fine, I have one."

"Right on."

Back upstairs, Dave turns a light on in the teacher's lounge. "Sleep in here on the couch if you want. There's a coffee pot and coffee. Don't worry about cleaning it up, we'll take care of that."

Am I dreaming? It's as if this superintendent has a wind-weary cyclist needing shelter every day. How can I thank him?

"Hey, I think I'm an honest person. You look like an honest person.

You're not going to steal anything."

"For sure, I wouldn't want to carry anything extra."

"Right on!"

We return to where we entered. "If you stay within fifty feet of this corner, your phone will work. It's a hot spot."

"My husband will thank you, I haven't been able to call home tonight."

Dave pulls out a business card and jots numbers on the back. "Here's the code for WiFi access. And my phone number. If you need anything, just call." We walk past a vending machine. "Need any food?"

"Nope. Got everything I need."

"Right on! If anyone says anything, just tell them I said you can be here."

Angels come in many guises. Tonight, Dave Marzolf, superintendent of Sun River Valley Schools, solved all my problems—food storage from grizzlies, protection from the storm, contact with home.

The rain hasn't hit yet, so I pack up and drag everything in. No sense taking chances. Alone in the eerie halls of Simms High School, my footsteps echo across the gym. I can almost hear ghostly dribbled basketballs, teen laughter, kids roughhousing. *Just my imagination.* I am relieved when I don't see the word "REDRUM" scrawled on the dank locker room wall. (Thank you, Stephen King.)

The shower's hot water softens my goosebumps. As I towel off, the Eagles' "Hotel California" overtakes Laurie Anderson. "Voices down the corridor…" *Wait. Those are voices, and this is not the Hotel California.* I dress in a hurry.

Light streams through open doors at the end of a murky hall opposite the gym. Silhouettes drag folding chairs into the school; two men stack chairs on long rolling racks. Outside, a third man in the bed of a pickup hands chairs to two young women.

"Hello?"

One stacking man startles and drops a chair. Everyone freezes.

"Dave Marzolf said I could stay in here. I'm riding my bike to Missoula."

The man in the truck breaks the freeze, handing down a chair. "For a cause or a charity?" His accent—are these folks Hutterites? Otherwise clean-shaven, the men's beards unfurl beneath their chins. Sandra-the-librarian said the Amish-like group is common in Montana.

"No, just for myself." I hang my head in self-absorption.

They turn back to work, so I join the women in hauling chairs. Turns out the Fort Shaw wedding party borrowed them from the school. One of the young women is in high school, she says she loves to ride her bike and thinks my trip is inspiring. "I would love to do that one day."

"Then you should. You'll learn you can do anything."

My first tours taught me this; eighty-four days across the country with B'76 reaffirmed it. We rode through snow, storms, wind, heat, over mountains, and across the plains. We pushed our bikes up Ozark hills, slipping in tar softened by heat and humidity. We shared cooking and clean-up chores, met people from all walks of life, survived crashes and dysentery. By the time I dipped my front wheel in the Atlantic Ocean, I was invincible. It's no different forty years later. My sixty-year-old body keeps ticking. I can do this. I *am* doing this. And I still meet amazing folks; my hope for humanity again renewed.

Perhaps if I inspire even one person, my ride is not so self-absorbed. "Believe in people, people," I write in tonight's Facebook postcard.

STATS
35 miles, max speed 21, ride time 3:23, avg speed 10.3, TOTAL 1804.62

Me and Superintendent Dave Marzolf.

32

Firecracker in a Calm Pond

Sunday, July 10, 2016
Simms to Augusta, Montana

Endless rows of tables stretched Escher-like across the smoky office. Keypunch operators, coiffed young women wearing dresses and high heels, sat stiff on straight-backed wooden chairs as if they balanced books on their heads. Crossed legs bounced in rhythm with the ticking of manicured nails on keys. The din was deafening.

One stop on an eleventh grade co-op career-exploring field trip.

"I may not know what I want to do," the girl thought, "but it isn't this."

A night on the teacher's lounge couch is heaven compared to sleeping through a storm on hard ground, yes? No. Abnormal gurgles from the turned off coffee pot, too-squishy cushions, and a motion activated light switch that beamed florescent rays with my every turn stole my sleep. When the ground feels more comfortable that a couch, I know my transition to life on the road is complete.

Lack of sleep gets me pedaling under clearing skies by 7:00 a.m. I look forward to today's short cycle to Augusta, the "heart of Montana," as named by yesterday's wiry man, despite having a population of only 300. He railed an imaginary lasso with his beer-free hand. "It's the start of the Rockies. It has everything. Fishing, hunting, farming, ranching. I went to the rodeo there and saw 100 people I know from around my territory!"

Bunny won't like the image of him that pops into my head, pissing his presence at every corner of his "territory."

Let it go.

A gradual climb along the Sun River presents nothing worth walking. Ahead, mountains rise over tumbling prairie grass checkered with irrigated fields. Three bunnies show themselves, two raptors, three deer (two heavy with velvet racks), and seven antelope. And herds of cattle. Billboard-sized metal signs, hanging between poles over driveways, feature CNC-cut silhouettes of mounted cowboys driving cattle with dogs chasing behind. Here: "Broken O Ranch Beef Division." Miles later: "Broken O Ranch Farm Division."

Ranch country.

My heart belongs to Michigan woods. But these vistas, life lived immediately, empties me. Truth—I am both wave and particle of a dangerous and beautiful nature.

A puff of dust chases me past Wagons West Hotel and RV Park into Augusta, the "Last Original Cow Town of the West." In my head, a tin drumbeat, a whistle, and a horn's WHA-WHA-WHA, subdues Laurie Anderson. The twang of a guitar and men's voices rumble the theme song from the movie *The Good, the Bad, and the Ugly*.

This can't be Augusta, it must be Rock Ridge, the fake town in a scene from Mel Brooks' *Blazing Saddles*, an illusion rising from the prairie. I half expect Main Street to be dirt, horses tied to hitching posts in front of this real-life town's Buckhorn Bar on this side, the Lazy B Bar and Café on that. Shouldn't buxom ladies lean over the picket rail on the upper porch of the Bunkhouse Inn? Tucked behind The Western Bar, Canyon Mountain Outfitters, and Latigo & Lace (all one-story weathered wood structures with two-story facades) is the American Legion Rodeo Grounds.

This must be Augusta. Would the smoky smell of bacon percolate from Mel's Diner if this were a movie set? The eatery is cozy busy: waitresses hustle plates and customers crowd for coffee. Outside the air is cool, inside is cooking.

I spread my maps across the table in a cramped corner booth. Time enough for second-breakfast even if I decide to push on. I pedaled strong this morning, yet I'm not sure about chancing sixty-some miles to the next camp in Lincoln. Dark clouds skip from the north, a rare tailwind if I continue, but the route is mostly up with an eight-mile climb over Rogers Pass (5610') forty miles in.

Today I rest. Wagons West RV Park it is.

The hotel office is closed, so I pick a tent spot next to a sheltered picnic table. The park is deserted except for a handful of RVs that appear rooted to the dry ground. More temporary homes for oil workers? My tent stakes bend. A chunk of leftover firewood makes for a good enough hammer. Now I'm a tourist for the day.

The screen door of Latigo & Lace squeaks as I enter, even as I try not to draw attention in my faded tights, yellow rain jacket, and green wind-charm still on my noggin'. The western-themed art gallery and gift store has a coffee bar. Definitely not Rock Ridge.

A tall, thin, middle-aged woman turns from organizing books, her floor length skirt flutters as if the wind blew in. "Can I help you find anything?"

"No, just came in to see what you have."

"Where're you riding?"

Well-versed in identifying a touring cyclist, I see. "Missoula." And before she can ask, I add, "from Michigan."

A woman who is browsing edges over. "Can I buy you a coffee? I'm originally from Detroit. I teach at a college in Connecticut now. I'm traveling to Glacier and taking a roundabout way back for the summer, stopping by Detroit to visit relatives."

Before I decline, she maneuvers closer. "I'm going to go get an ice cream. Come with me, my treat!"

Her energy is like a firecracker exploding in a calm pond. I step back with a sudden longing for the quiet of the empty prairie. "No thanks."

She bustles to pay for a pack of hand painted postcards; I take advantage of her absence and head for the door. "Good luck on your trip."

I cross to Allen's Manix Store, AKA The Trading Post, AKA The General Store. "IF WE DON'T HAVE IT, YOU DON'T NEED IT!" screams in faded paint across its upper story. The sign is right, the place overflows with fruits and vegetables, dry goods, housewares, camping, fishing, and hunting supplies, souvenirs, cowboy hats (I am so tempted), automotive oil and parts, tools, and kids' toys. Less than $10 buys me an apple, banana, avocados, sharp cheddar cheese, flour tortillas, and a one-size-fits-all pair of tiny, knit gloves (my hands are cold).

The office is not yet open when I return to Wagons West. I take a seat outside on a slat bench to FaceTime Mom. Hard to see her so frail, slumped on a heating pad in her Lazy-boy.

"My back hurts and I'm just tired. I'm sad about Ag."

Something else bothers her. She blurts, "Cathy and I had a fight."

Cathy. My sensitive artist sister who visits Mom every week, takes her shopping, to the doctor's, or the bank; who struggles to let Mom's acrid remarks roll off her back; who continues to visit, even now that Dad is gone, because she doesn't want regrets.

Mom sobs her story. "I don't know why I'm telling you this, I wasn't going to."

"Because you can talk to me." And a long history of listening: months while Dad commuted to Clare when I was in ninth grade at the public school; the following summer when Dad quit his job; more months of him commuting a return to the city. It doesn't take much, and I am back on the couch with her, watching *General Hospital* after school.

"I've said some things I'm sorry about. Dad always said, 'Be careful what you say because it's always there.' Cathy sent me a long email. I'd be okay if she'd just leave me alone, but she keeps calling."

I keep listening.

Poor Cathy. Her calls come at bad times, according to Mom, when she is going down for lunch, or about to take a shower. Cathy says, "I'll call back," but when she does Mom is going down for dinner. I can almost taste the tartness in Mom's voice, "I'm sorry Cathy, I'm busy."

Mom chokes "love you" and throws a kiss. "I can't wait until you get home," she gurgles as the screen goes dead.

I stew. Am I selfish for wanting to keep riding? Ever since I can remember, I felt called to care for my parents when they aged. I didn't expect that their hard independence could feel like rejection. So, what did I do? I moved three hours north. Far enough from the city, close enough when they needed me.

I know I have to live my own life. I love my life with Andy and moving to the northern woods is the best choice we've ever made. Now here I am, chasing a personal grail, thousands of miles away.

Bunny snuffles. *Didn't you ever consider they wanted to live their own lives too?*

STATS
24.15 miles, max speed 31.5, ride time 2:18, avg speed 10.4, TOTAL 1828,81

Mom holding me in 1956, surrounded by Cathy (left), Rick, and Sue (right).

33

Unexpected Counsel

Monday, July 11, 2016
Augusta to Lincoln, Montana

Hot kisses, hot hands. High school nights parked in the driveway in his red MG, steaming up windows.

The girl snuck into the dark house, flush with lust. Her mother caught her at the bathroom door to whisper a warning. "If you feel that way about a boy, you'd better marry him."

The girl wiped her mouth, stifling a retort. "But then I'd have to marry them all!"

Raindrops rapping on my tent brings to mind the song "Rain on the Roof" by the Lovin' Spoonful. A silent apparition connects me to the past—in one of Dad's home movies, Grandma Brehler sat down at her piano at the exact moment when "Rain on the Roof" burst from the radio. Her wailing up and down the keyboard was a perfect mash and as the song trailed off, she gave a final hammer and turned to the camera with a seductive smile. It's rare I can hear this song without remembering her smile.

This morning, it's me and only me caught up in a summer shower. I doze. I read. I call Andy. I dig out a tortilla and eat an avocado and tomato wrap for breakfast, disassemble my pillow (sleeping bag stuff sack filled with spare clothes and jacket), stow sleeping bag and pad, and repack my dresser pannier. Organized for the ready, I can delay until at least 11:00 a.m. and still make Lincoln before dark.

"Hello!" Déjà vu, man's voice outside my tent.

I stick my head out and come face to face with a German Shepherd, who schmags my glasses with a gentle lick. Beyond the dog are paint-spattered jeans attached to a tall man with a Santa Claus beard and a cowboy hat.

"Good morning! We were out picking yesterday and thought you might like some cherries." He extends a Ziplock full of yellow-to-orange-to-red-toned Rainiers. I crawl out to thank him.

"We're living the dream. Building a log house on a chunk of land over those mountains." JP and his wife, Jane, are staying in one of the rooted RVs here until it's done. "Where're you riding to and from?"

"I started in Michigan on June 9, heading to Missoula."

"I'm originally from Michigan!"

"Where?" When he answers Clare, I'm not surprised. Coincidences follow me like flies.

"You probably don't know where that is," he adds.

"But I do. I lived in Clare for a while when I was in high school." Dad took a job transfer and we moved north at the end of my freshman year. Rick was long out of the house, Sue was married with a baby boy, and Cathy (a year out of school) decided to stay in the city.

He slaps his thigh. "When? What year did you graduate?"

"1974."

"Me too!"

We don't remember each other, but we know many of the same kids and teachers. There is no doubt. We were classmates at Clare High School in 1971-72.

"You need to come over here and meet my wife Jane. Jane, come out here!"

A lean woman with a shock of hair as white as JP's beard opens the door. "Come on in, it's still raining."

"Can you believe it? Patti and I went to high school together in Clare."

Second breakfast, second day in a row. Jane pours me a healthy cup of steaming coffee and a bowl of cereal loaded with nuts and bananas, served with a hunk of hearty conversation. We catch up on years we never realized we missed.

JP was raised Catholic, yet like me, is no longer. He is one of eight kids, I am the middle of seven. He holds multiple degrees, was a social worker, wilderness/park ranger, and college professor, recently retired from Minnesota State University, Mankato. He is also a dowser, eager to meet with a well-digger on his land this week.

I dropped out of college more than once. Worked in men's fields most of my life. Factory rat, journeyman machinist, massage therapist, bike shop owner and mechanic, repair technician and dog trainer, freelance writer and editor.

Jane is a retired counselor. Her first husband, a doctor, is deceased. He had three adopted adult children when she married him; she had no children of her own. JP has two kids with a twenty-year age span.

Talk turns to today's world.

"Rural people are the salt of the earth," JP asserts.

I cannot disagree and share stories of kindness during my travels. "It makes me sad that my experiences might be different if I were a woman of color and not privilege."

The couple nod in agreement even as JP offers a hint of hope. "I find people here accepting one-on-one, even if fearful of the 'other.'"

We share inner journeys. JP had many, I am on one now. I confess my conflict about taking the train or riding home. "After talking to my mom last night, I wonder if I should take the train. Andy would like that, maybe I should think of them."

"'Shoulds' always concern me," Jane jumps in. "What do *you* want to do?"

Ride home.

JP adds two cents, "I expect your mom's feelings will be what tempts you away from what you want."

I think he is right.

Reluctant to leave, it is after my 11:00 deadline. We take photos, share contact information, and promise to keep in touch. Before heading out I FaceTime Mom, she is feeling better.

I am revitalized. A side-to-slight-tailwind invigorates a 14.2 miles per hour average speed for the first eighteen miles out of Augusta.

A turn and headwinds return. Growing headwinds. Three dreadful hills present 1200 feet of elevation gain (2600 steps pushing) as the road switchbacks up to the Continental Divide. Craggy slopes squeeze the partly cloudy sky to insignificance. There are times I barely make ground against the wall of wind, I long for an oxen yoke to help lean into my load.

Tips of pines hang below and soar above the narrow ribbon of road etched against the mountainside. Yet another zag and the road widens with gravel pull-offs on either side. Around the next zig, a compact car cruises toward me and slows. A metal railing doesn't look strong enough to hold it from a deadly drop if its brakes fail. Two young women dressed in shorts and tank tops jump out. One edges to the guardrail to photograph the descending valley, a magnificent and inspiring view.

I pause in my pushing to breathe. The women wave. "Where're you riding to, where're you from?"

The wind muffles my answer. My voice is hoarse. I realize I'm shouting. "I've had headwinds almost all the way. I'll be riding back too." I am frantic in my telling, as if I haven't spoken to another person in weeks. "How far to the top?"

One looks back. "Maybe ten miles?"

"NO!" *It can't be that far, two or three maybe, not ten.* I deflate into my handlebar, round the next curve, and remount to pedal what I can. Wait. What's that green sign?

"They're wrong!" I holler at the wind. Rogers Pass, elevation 5610.

The allure of a slow slog up a mountain is the other side. I relax into growing G-forces as the road curves a delicious descent. *What was that marker I zipped by?* The coldest recorded temperature in the continental US (minus seventy degrees). I am glad for tiny knit gloves.

Abruptly the road climbs again. *What?!* Trickster mountain, making more work before a roller-coaster descent. At last my Tour Easy swoops sure, its weight a rolling advantage. Ian Anderson's flute, in Jethro Tull's "Fat Man," chases Laurie from my head and lyrics about racing a fat man down the mountain makes me smile. With every riff, with every curve, the mountain spreads its arms open like a lover, vista expanding as I return to earth. Gone is the dry sweeping prairie. Here are forested hills and deep green pastures.

Above, an eagle circles, chased by a little black bird.

The last twenty miles into Lincoln is as easy as the climb up Rogers Pass was hard. Mountain ranges and Ponderosa Pines ring this town of 1000 people. Its main street is littered with shabby hotels and scruffy bars, each with a mini-casino. Until Mikey told me, "The only thing I remember about Montana is all the casinos everywhere," I never noticed. Slot machines are as common as vending machines; even the D & D Foodtown Grocery has one.

"Is there a laundromat here?" I ask the cashier.

"Over there." She points out the window to a mobile home park one block west, tucked behind Lambkin's Casino Restaurant and Lounge.

Weathered homes are stacked like dominoes in a park not meant for travelers. Here a screen door hangs on one hinge, there a young man is bent over the engine of a rusty pickup truck, a cigarette hanging from his lips. I weave my way around the puddle-pocked dirt street. The laundromat is a ramshackle shed. Inside, half the muddy street is on the tile floor, no change or soap vending machines. *I can wear this outfit another day.*

Maybe it's the threatening sky, or dusk coming early, or neglected buildings without people out and about—Lincoln is depressing. I cruise a hardscrabble neighborhood looking for a second choice to camp. Not inclined to backtrack to Hooper Park, a rest-stop-like, self-serve campground I passed on the way into town, I keep west.

Behind low brush and pines, a sign peeks. "Spring Creek RV Just Ahead." Three trailers occupy half the RV sites. No one is around. A restroom building sign welcomes bicyclists with an eight-dollar special: tent camping in a meadow, free shower and WiFi, self-registration.

Sold. Sanctuary under a gnarled old tree with a chittering creek for company.

STATS
59.56 miles, max speed 39.5, ride time 5:57, avg speed 10 (was 14.2 the first 18 miles), TOTAL 1888.40. ROGERS PASS 5610 WINDY!

JP, me, and Jane.

34

Stranger in Town

Tuesday, July 12, 2016
Lincoln to Ovando, Montana

The girl, one of seven valedictorians in the senior class, stood firm with her back against the wall, the football coach towering over with clenched teeth. He was the school's graduation coordinator. "I don't understand you."

"Why? Because I won't walk across the stage for a phony diploma?" They held the actual diplomas until the return of caps and gowns.

"You have to give a speech."

"You don't want me to do that," she warned, rehearsing rebel thoughts critical of an institution she believed did little to prepare self-absorbed students for "real" life.

"And you haven't even applied to college."

"I don't want to 'be' anything. I only need a job that pays enough to support my habits."

The sound of his fist punching his palm echoed through the hallway.

She stood a little taller. "I just want to get on my bike and ride."

In Oregon on June 2, 1976, snowpack on McKenzie Pass closed the road across the Cascade Mountains until July, forcing my B'76 group over the alternate Santiam Pass. Even Santiam was chancy. Rain at lower altitudes turned to snow higher up. We took refuge in a steamy laundromat until plows cleared the road. At the crest, some of us stopped for an impromptu snowball fight. The California-born tandem couple had never seen snow. For this used-to-snow Midwesterner, what struck me as uncanny was the abrupt disappearance of the pine rainforest as we crossed the volcanic range—red and orange striated rock bluffs, dotted with sagebrush, dominated the eastern high-desert landscape. That's how 130 inches of precipitation a year on the west side compares to fewer than twenty inches on the east.

Rumor was a sixty-something "independent" (unattached to a group) rider pushed his bike up McKenzie through thigh deep snow and was forced to camp there for the night. I believed the rumor. The tough old geezer rode along with us for a few weeks. His bulging belly, white beard, and bluest, truest sparkling eyes reminded me of Santa Claus. Every day he hit the road before we crawled out of our tents. We'd pass him later, but he was always last to bed, often after closing the local tavern.

That was early June. This morning, in almost the middle of July, the thermometer on my handlebar bag reads thirty-two degrees. No snow, but not the temperature I expected, even at 4500 feet elevation. I might be a solo-riding, tough old geezer myself now, but I don't close out the bars. And in the morning, I sometimes linger.

The road ripples out of Lincoln. Riding west of the Continental Divide with the wind in my face yet again, I am not sorry to leave the plains behind. My bandana wind-charm makes the fight tolerable. Or maybe I'm getting stronger. The sun soon coaxes off my tights.

A semi pulling two trailers loaded with wood posts roars from the west, leaving the sweet smell of fresh-cut lumber (and a pang of homesickness) in its wake. JP and his wife might live the dream here in Montana, but Andy and I live our own dream in northeastern Michigan woods.

"I never thought you'd leave your family," Andy said when we moved out of the city. "They'll think I knocked you over the head and dragged you north."

"Oh, please don't throw me in that there briar patch!" I countered.

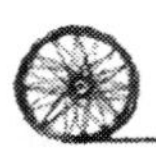

Andy knew of my commitment to my folks, so he never broached his yearning to live in the country. He never knew I'd consider it. Funny how it came about.

My brother, Jim, and his wife had a vacation cabin in northeastern Michigan, but after he quit his high-paying engineering job to follow a calling into the ministry (and a move to Ohio), they didn't use it much.

He couldn't sell because of the real estate bust, so Andy and I helped with the mortgage. We loved going north often.

When Jim eventually sold it, we spent some weeks looking to buy our own little cabin. One Sunday, we passed a log house I had always had my eye on—the last private land next to the Huron National Forest, near the banks of the AuSable River.

"Andy, it's for sale!" Vacant and needing work, it sat on twenty-six acres with a detached garage and horse barn.

"I don't think we can afford this, and our townhouse," Andy said.

I stood on the covered porch, peering in a bring-the-outside-in window of a corner room, and envisioned myself writing there. I turned to him. "I could live the rest of my life here."

"You mean you'd move here?" The look on his face was priceless. "What about your parents?"

"It's only three hours away, close enough go down anytime I need."

Monday morning Andy had a pre-approved mortgage; by Wednesday we presented an offer. But my dream house was not to be, we were outbid. I became obsessed with getting out of the city, and not long after we found our patch, which turned out to be a better place. Move in ready, with thirteen acres of hardwoods abutting more than 4000 acres of state land. The separate twenty-four by thirty-six-foot, heated workshop sealed the deal for Andy.

I doubt Mom was happy at our leaving. Ah, well.

The Blackfoot River meanders with SR 200 leading west. In scenes straight out of the movie *A River Runs Through It*, fly fishers wade in glistening waters, their arms a pendulum tick.

I am but a sparkle, floating on the man-made winding, the road built next to the river. My eyes rise. A steep green slope rears into blueness, where pine trees pierce puffy white clouds. I wrench my eyes from dizziness. At every turn, the vista widens; ahead, the Garnett Mountain range dominates the horizon. Yep, snow up there.

A shadow swoops and I can't keep my eyes from lifting again. A raptor soars above an opening valley. Along a ridge, two more stand guard on fence posts. With a screech, the two birds take flight.

A transcendental ride to Ovando.

A lone sign on SR 200 assures passersby, in circus-red letters, "OVANDO IS OPEN." I swing by Trixie's Bar at the corner to roll into the town's triangle, ringed by the Ovando Inn and Blackfoot Commercial Property on one leg, the Blackfoot Angler and Stray Bullet Cafe on the second, and the Brand Bar Museum on the third. Street names are chiseled into rough-stained, one by four boards screwed onto cedar posts, with metal silhouettes, much like ranch signs, mounted on each. The corner of Short and Pine Streets sports a cowboy on a horse with a pack mule tied behind.

Ovando. "Jewell of the Blackfoot Valley, POP: about fifty, Elevation 4100, DOGS: over 100." Not only a stop on the ACA Lewis & Clark Trail, Ovando is also on the ACA Great Divide Mountain Bike Route. This off-road route crisscrosses the Continental Divide from Banff, Canada to the Mexican border in New Mexico.

Cyclists sleep free on the grass in front of the museum, or, for a $5 donation, protected in a hoosegow, tepee, or shepherd's wagon. I called "Kathy S." at the Blackfoot Angler last night and made reservations.

"I'll pin your name on the door of the hoosegow," she said. "It's all yours."

Kathy S. was full of stories about the Great Divide Bicycle Race on the ACA route. The self-supported, non-stop, off-road event started in Banff on June 6 this year with about 150 racers. The winner set a record with a time of less than fourteen days.

Yowzers.

"We got in trouble from Adventure Cycling this year. One woman dragged into town so cold she had a hard time peeling her hands from the handlebars. The racers aren't supposed to get help from anyone, but here in Ovando, if we see someone who needs help, we're gonna help them."

I don't doubt her.

In the hoosegow, a Deer Lodge County Jail built in 1890, thick planks separate a guard room from a cell. A rustic table and chair sit under a barred window; two rope beds in the cell hang from hand-hewn walls. Except for the lone window and shards of light beaming where chinking has worn away, all is dim. A sudden squall thunders in, pelting rain through the cracks, and then thunders out. I am grateful for the heavy beams and drag my bike and gear inside.

The landmark Ovando Inn and Blackfoot Commercial Company houses a combination of services—convenience store, hotel, espresso bar, one-pump gas station, and laundry. *I can get my clothes washed!*

The owner meets me at a side door. "We don't have a dryer." A bedroom-sized room houses a lone washing machine. "Well, I bought one used, but it broke. I've a new one coming in. You're staying at the hoosegow? Your stuff should dry if you hang 'em up."

With clothes in the washer I sit on the stoop to write. Something buzzes me. I jerk and swat empty air. A green, yellow, and orange hummingbird hovers a few feet from my face as if checking out the newcomer in town. "Hello." I smile. The bird zooms across the road to survey the hoosegow.

I miss Bunny. No sightings in a few days. "Did you send the hummingbird to check on me? Today's transcendental ride along the Blackfoot River might have pleased you, Bunny."

STATS
27.45 miles, max speed 30, ride time 2:14, avg speed 12.2, TOTAL 1915.91

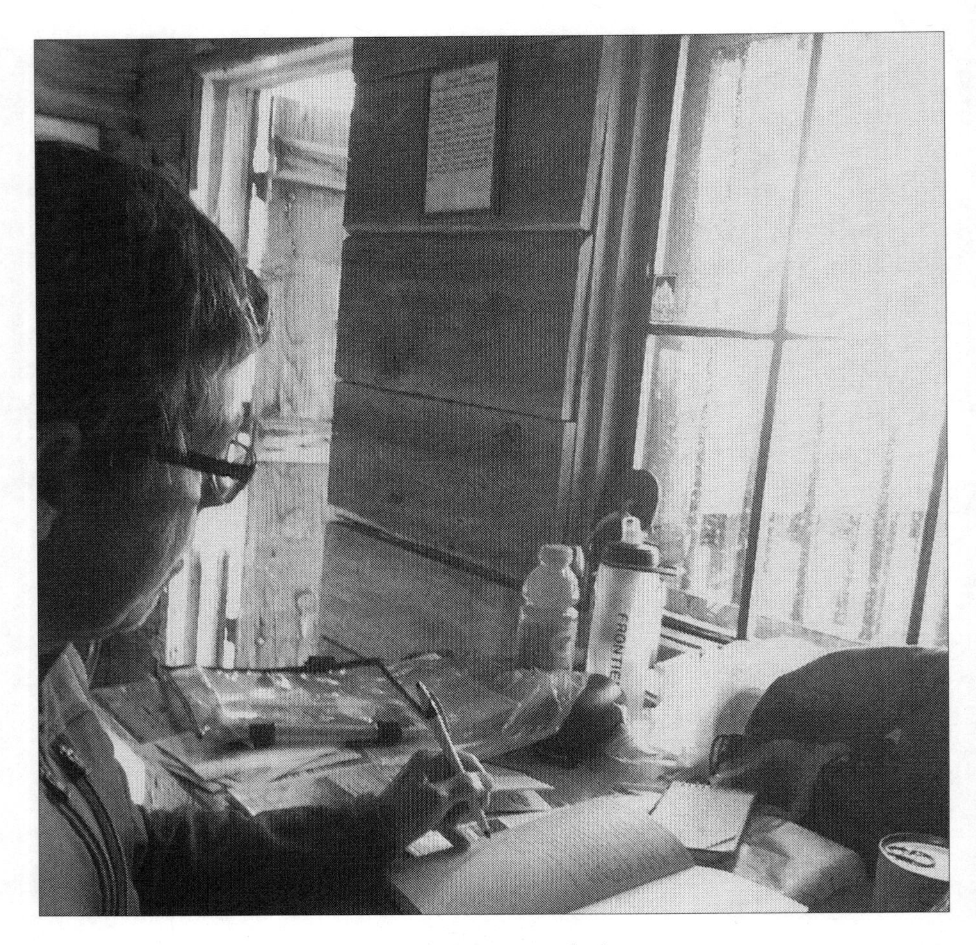

Journaling in the Ovando hoosegow.

35

Soldiers of the Sorrowful River

Wednesday, July 13, 2016
Ovando to Potomac, Montana

The family moved to a rented farmhouse the girl's junior year. New roads to explore on her Sears single-speed, but hills gave her a ten-speed craving. By the end of summer, the girl saved enough money from her job in the penny-candy store. Her father took her shopping.

Tippy-toed over the crossbar of the first bike the salesman showed her, "I want this one."

"It's too big," her father said. "You need to find one that fits you."

He was right.

Another move her senior year, closer to the city and a different bike store. This salesman put her over a Raleigh Super Course; it fit as close as possible in the days before women-specific designs. She recognized the bike's Simplex components from the bicycle touring book Aunt Mary gave her. Best of all, it sported a Brooks leather saddle.

Sold.

Oh, the glorious odor of bacon grease and coffee. Like a bird dog trembling at the scent of grouse, I point myself to the Stray Bullet Cafe. Vintage photographs of the town and neighboring ranches hang on log walls. An eclectic mix of bentwood and square back chairs, arranged with round and rectangle wood tables, dot the faded tile floor. Each place setting is ready with a hand-thrown mug. Rain sprinkling on the out-sized front windows encourage me to dawdle. Ovando locals and visiting anglers trail in to load up. A sudden pang. I miss my weekly breakfast out with Andy at the Sunrise Café, where every morning the local "linger-longers" share bites with a bit of gossip.

One man, I can't tell if he's a guide, or a fisherman headed to the Blackfoot, picks up a box lunch at the counter. He stops at my table.

"Hey, I'm Dain. Came up from Livingston to fish. Where ya from? And where ya headed?"

Again: Michigan, Missoula, the anniversary celebration.

"Awesome. When my friend and I retired our wives said, 'You two need to do something.' So we got bikes. But they don't want us to be gone more than twenty days. We've been riding the Lewis & Clark Route in sections." The Lewis & Clark goes from Seaside, Oregon to Hartford, Illinois. Dain keeps on, "This fall we're continuing south. We're trying to convince our wives to meet us in Key West. Then we'll have done a diagonal and I'll be done."

By connecting the coasts, Dain and his buddy will ride a "true" transcontinental. I say, "I'm thinking of taking Lewis & Clark eastbound through southern Montana on my way back."

"Do you have some paper and a pen?" Dain writes his address and phone number. "If you return along the Yellowstone River call me. You'll go right through my town."

"Thanks, and good luck. Even if you make it to Key West, you may find you'll never be done." *Just look at me.*

All this talk of riding makes me antsy. These recent low-mileage days must be the rest I've needed. I'm eager to turn my rig home, eager to ride hard. Bunny sniffs. *"Patience, little one."* I disregard him. I don't care if my clean and dry clothes get wet, I gather my things and brave the rain.

Rain dries into a lovely, if overcast, morning. A bald eagle solos above the mountains and raises my heart. I could do this forever. I know I cannot.

Fifteen miles into the day's ride, I stop at a gas station/convenience store where Highway 83 from Glacier National Park intersects with SR 200. Second breakfast of yogurt and a chocolate Power Bar. I can't believe they still make the grainy things, a staple in my racing days.

I take my place behind three women in line for the outside-entrance restroom. A man rides up on a loaded, lugged steel Raleigh touring bike, dating back quite a few years. Jack is from Kalamazoo, a fellow Michigander. He took the Amtrak from Chicago to White Fish, Montana and is riding to Missoula for the celebration. His wife is flying out to meet him. She rode part of B'76 but left the ride in Kansas to work for the organization.

"I plan to ride to the west coast to see my daughter after the party," he says.

Something nudges me away. *Bunny?* "Have a good ride. I'm going across the road to the rest-stop, this line isn't moving."

On the grass outside the no-lines rest-stop bathroom, a woman waits out her sniffing dog. The pup is a lean, fewer-than-forty-pounds, black-lab mix and reminds me of my old mutt, Gypsy.

"Can I pet your dog?"

"Sure. This is Lolo, named after the pass between Idaho and Montana."

"Hey Lolo, I rode over that pass forty years ago."

Joanne is fifty-seven years old, a civil engineer retired from the city of San Diego. "I worked on bike paths. Sold my house in three days and bought that." She points to a small trailer hooked to a dusty maroon pickup. She travels wherever her heart takes her. "I'm grieving the loss of my soul-mate dog, Bo. Didn't realize how much unresolved grief I've been carrying around. My brother died several years ago. He was forty-seven."

Tears well. I hug her like a momma bear.

Until sheets of rain sting us, we don't notice the storm blow in. We duck for cover under an overhang, shuffling to let travelers pass. I share my grief from Dad's passing. We agree, grief compounds. We might learn to live with it, but sadness hangs heavy.

Together we wait it out.

"There," I point. A patch of blue-sky peaks behind clouds no longer desperate and puffy, like our eyes. Joanne and I, soldiers of the sorrowful river, hug goodbye to carry on.

I pedal. The road flows again with the Blackfoot River, rushing in rhythm with my mind. Another eagle. I imagine being held in its talons, unafraid, watching earthly cares retreat down a column of air.

Free and not alone.

Tonight's destination, a bed-and-breakfast with free camping for cyclists, is four turns and fewer than four miles off-route. Potomac Road winds into a green valley of horse farms. A bank of mailboxes marks the second turn onto an unmarked, pot-holed road. I creep along. *Bunny, is this the right way?*

A four-wheel-drive truck pulling a horse trailer bangs up from the opposite direction. Two women are in the front seat. I wave them down.

"I'm trying to find Sundog Permastead. Am I on the right track?"

"You're doing good," the driver says. A panting lab edges to greet me from the back seat of the extended cab. "Keep going. You'll wind through the trees and out of them. Don't take the road up the mountain. Stay on the pavement. Then, when the pavement ends, turn left. It's uphill, I have to tell you."

I shrug. She adds, "But it's not that bad."

The ride through the valley is worth any miles off route. Less alien to me than the wide-open prairies, hillsides of towering Ponderosa Pines diminishes Montana's big sky. The gravel uphill is bad enough. Ah, well, I walk much of it. Two antlered deer bound across, hooves clattering on the rocks, and disappear. A final sharp turn rises into a rivulet-ed dirt drive. I lean into my load until the ground levels.

A twelve-foot pine slab, wedged between two trees on one end and resting on a stump on the other, displays a dancing dog and the words, "Welcome to Sundog Ecovillage." Other wood signs point the way to "Yurt Cottage," "Camping," "Cabin," and "Park." Behind a stick fence sprawls a one-story house with a steel roof.

Have I time-traveled to the 1960s?

Two barking dogs bound over, but the gray-faced mutts are more welcoming committee than guards. The furrier, shepherd-mix with a huge grin wiggles over, eager for a scratch. The black and gray mottled one with a curled-up tail keeps his distance.

No other sign of life.

Yesterday, I called Hunter, the owner, to reserve a campsite. He said he might not be here when I arrive, so I wheel past the yurt and several sheds to the camping area. The ground is soft with pine needles. Four logs square off a tent-site graded with mulch and a solar-powered lantern is stuck in the ground at one corner. Next to it on a sheet of plywood is a metal table with two logs on end as chairs; beyond are two metal lawn chairs and matching end table.

A compact car pulls in. Don is a friend of Julie's, co-owner of Sundog and mother of Hunter. He welcomes me and invites me to the house to meet Julie, who obviously didn't hear her greeting committee.

"Hunter didn't tell me you were coming," Julie says.

"He said he'd leave me water, but I didn't see any."

Julie grabs a gallon glass jug of water and walks me out to the site. The pups join us. Camus, the shepherd-mix, came with the property when Julie and Hunter bought it four years ago. Earl started out as her dog then went to another son. When the son joined the Marines, Earl lived with her daughter. "Now he's mine again. He has seizures. He's not all there."

Camus drops a ragged ball at my feet. I am happy to throw it. And throw it again. And again.

Julie points out the composting outhouse, a cross between a stick tepee and a pyramid. Six steep wooden steps lead up to the door, two long windows drape the sidewalls. "Don and I have to pick up my mother from the airport. She's visiting from Ohio. Let yourself into the house to shower."

"Thanks!"

"There's cell-service on the porch, but not out here. We're expecting two guests in one of the rustic cabins. They'll be in after we get back, so don't worry."

Camus and Earl give me a tour of their mountaintop domain. We visit the open-sided tent where Julie serves breakfast for paying guests. I peek into the yurt and cabins, welcoming with colorful quilts draped over cozy beds and writing tables positioned at low windows. Who would have thought an outhouse could be so luxurious, with magazines, incense, and a box of sand to layer after use?

Ah, the view.

Pines stretch heavenward, their topmost branches dance with the wind. On the ground, where mere humans stroll, the breeze is gentle, and cool when clouds obscure the sun. Insects drone a symphony across scrub-grasses. A pause. The silence left behind renders an implosion in my head; before I succumb, the drone returns.

Where sunshine reaches the ground, wire-fenced gardens burst with herbs. A young deer, stuck inside one, darts from end to end. It launches free at a sag in the fence and turns to stare.

We wander. I take photos of wee purple wildflowers with yellow centers. The dogs escort me to a clearing with an organic Stonehenge of two-foot diameter log ends ringing a fire pit. I imagine sparks rising to the Milky Way, a bridge between earth and the cosmos.

A clear night will reaffirm my insignificance.

The waning afternoon makes for pleasant reading. Stegner's story-in-a-story is about a wheelchair bound, retired history professor writing his grandmother's biography. I'm hooked. Or maybe it's the independent (and ahead-of-her-time) grandmother he depicts. The character is Susan Burling, an artist and writer born in the east who followed her mining engineer husband west. I empathize with her struggle with (and eventual acceptance, perhaps even love of) western prairie landscapes. Burling loves her husband, even as she harbors a close relationship with a lifelong woman friend.

The friendship Stegner describes reminds me of the lifetime I've shared with Lou. We met on a twenty-mile breakfast ride the summer of 1980. Her comment, "I used to run this route," caught my attention. She inspired me to run. The first time we ran together, we showed up wearing identical t-shirts. Tandem riding seemed the next logical step.

I joke that Lou is my "evil twin" with crazy ideas. "Hey, let's do the Paris-Brest-Paris," or "What about this RAAM qualifier?" Or "Let's do the Run Up the Empire State Building."

Every time, I said, "Okay."

Over the years, despite sometimes being geographically separated, we've remained closer than sisters. Like Burling, we persist.

STATS
33.18 miles, max speed 37, ride time 2:54, avg speed 11.4, TOTAL 1949.19

36

Nudged to Share

Thursday, July 14, 2016
Potomac to East Missoula, Montana

What the girl didn't know before a month-long bicycle tour around Michigan right after high school:

- *That a single-walled $25 Kmart tent will drench your face in the morning when you sit up.*
- *That cooking breakfast with a Sterno can won't result in crispy bacon.*
- *That rotten duck eggs left in campfire coals don't burn, they explode.*
- *That if you are going to illegally camp, go deeper into the woods.*
- *That if you don't call home for a week, your mom will freak.*

Stars that look like a fireworks finale explode in the treetops at 4:00 a.m., making short work of ego and long work of a kinked neck. I've come a long way in thirty-five days. Today I stop within spitting distance of Missoula, partly to save a night's hotel fee, partly to meet the niece of a woman I hope to get to know better, mostly to keep to my life on the road.

I dreamed the B'76 anniversary dinner was formal. I was in the stands of a football field and someone from Ogemaw County brought me a blouse to wear. I was angry about having to get dressed up. Telltale. The celebration was an excuse, not a goal. But I wonder. *Why did I need an excuse?*

Julie and Hunter, bundled in sweats and mud boots, come to hug goodbye. The dogs aren't with them.

"Did you find Earl?" I ask. He wandered off after our tour yesterday.

"He straggled home sometime last night," Julie says.

"That's a relief."

She checks out my Tour Easy. "This is what you're riding?"

"Go ahead, sit on it."

She leans back, hands on the handlebars, and smiles. "Comfy!"

I feel a kinship with this pioneer woman, we could be friends if I was a neighbor instead of a gypsy.

Hunter tells me about the natural healing salves and cosmetics he makes. "Do you need anything?"

"No, I don't use cosmetics."

"I make a balm to dress sores. Could you use that?"

It seems polite to purchase something for their generosity. "I suppose. I had chafing early on, maybe it would've helped." I worried about that so long ago, I gave up finding another pair of baggy shorts.

Hunter runs to the house and returns with a small jar. When I ask how much, he says, "Nothing. It's a gift. You are only the second biker to stop here. The first was a man several years ago."

"Thank you. You probably don't see many because you're so close to Missoula. Or maybe because you are off-route. I'll tell other riders about you. This is a beautiful place to stop."

Serenity blankets the valley out of Sundog. Horses graze behind split rails, a town of Columbian ground squirrels skitter between mounds, three athletic ranch dogs give noisy chase.

"Mornin', pups! Way to make sure I keep going."

Back on route the ride is splendid. Puffs of clouds dot the blue sky and the air smells of pines after a rain. The Blackfoot River curves in tandem with the road and teases me to consider fly-fishing. I see a fellow cyclist ahead. Jack. I slow next to him on the shoulder that is wide as a full lane. His rear derailleur pulleys have a bad case of chirping bird syndrome.

"Good morning, Jack. Did you get caught in the rain yesterday?"

"No." He keeps his attention ahead.

"I stayed at the most amazing place. A cyclists' only camp a few miles off-route. Isn't this a beautiful day?"

"I had a terrible night sleeping on rocks."

Jack camped at the Lubrecht Experimental Forest Station about ten miles back. I dare not tell him about the soft mulch I slept on. We ride side by side. Jack is quiet with a keen eye out for traffic. There is none.

"I'll give you the lane," I say. "Have a good ride, maybe see you in Missoula."

Some miles later the hillsides open to an extended view of the river and road bending a long, gentle "S."

"A good photo op, yes Bunny?"

I enjoy the wildness for some time before the neon yellow of Jack's jersey pops between the trees. Click, click, click, I zoom in as Jack pedals by. His forearm muscles bulge, his hands grip his bars like talons. When he sees my camera, he breaks a grin.

Erika's house in East Missoula sits on a hill in a modest suburban neighborhood. Her front yard bursts a garden green with tomatoes, Brussel sprouts, squash, peppers, and a row of sunflowers.

"Patti! Come on in." Erika opens wide her door and arms, then breaks away. "You've got a tick behind your ear!" She circles for a better look, guiding my fingers behind my right ear.

"Can you pull it off?" Funny how even this third tick hasn't caused me to flinch. *Am I doing ok, Bunny?* Must have picked it up yesterday walking around Sundog with Camus and Earl after my shower. I wore tights but neglected to spray repellent.

"Wait, let me get some peppermint oil."

"What does that do?"

"It's supposed to kill them. Give it a minute, its legs are still wiggling."

"Please, take it off."

"Are you sure?"

"Absolutely."

Erika extracts it, head intact. "Do you want to save it?"

"Ah. Nah." Silly me, I have no idea why I'd want to save it.

Erika gives me a tour, pointing out where I can take a shower. We peek into her husband's office to say hello, it is full of guitars and ukuleles. Shane works from home on computer stuff about which I don't have a clue. Erika is a birth doula. I meet sons Gus (nine) and Simon (eleven) playing on a screened-in trampoline in the backyard. They've positioned a sprinkler underneath, a brilliant way to keep cool.

"You'll have to fight the chickens for a spot in the yard for your tent," Erika apologizes.

During our Facebook messaging before my arrival, I learned Erika expects two friends for the weekend. I suggested I cook pasties for all of them, a UP Michigan treat Mom taught me to make.

"What do you need for dinner? I might have stuff, but I need to go to the grocery store myself. The kids just got home after six days at Grandma's."

"I'll buy what I need."

The urban, stocked-with-too-many-choices grocery is overwhelming. I'm guessing Erika will shop awhile, but halfway through aisle seven my phone rings.

"Hey Patti, it's Erika. I'm finished and have taken my stuff out."

"I'm still wandering the aisles."

"Take your time. Just wanted you to know I'm walking down to the hardware in case you come out and can't find me."

I beat Erika back to the car. "I'm sorry, just so much to look at!"

"Boy, if I knew you'd be so impressed, I would've taken you to our good store."

Chopping potatoes, carrots, and onions on a kitchen counter with Erika preparing a salad opposite is a rare party. We jam to the music Shane throws from his smart phone through a customized sound system, while he plays a board game with the boys.

Something nudges me (Bunny?) to share. "It's a crazy thing, but during this whole trip I've had a song in my head all the time."

"What song?" Shane asks.

"'Ramon,' by Laurie Anderson."

He picks up his phone and the familiar drum rift synchronizes with the beat in my brain. Hearing "Ramon" outside of my head is gratifying.

"I like it," Shane says after a few moments. "Reminds me of Kate Bush."

Erika's friends arrive and we take the feast outside. Gus and Simon surprise me with their enthusiasm for the pasties. "We love them!" Turns out Montana has its own history with the cuisine. The savory baked pastry was a mainstay for miners in the 1800s—easy to carry in pockets wrapped in newspaper to keep warm.

The evening wanes. We are all full. The boys and the friend of Erika's friend finish playing catch, and everyone goes inside except Shane and me. Talk turns to music.

He asks, "So why 'Ramon?' Why that song in your head?"

"Well, it's a great song. I first heard Laurie Anderson back about 1990."

I pause.

"My brother's wife was dying of breast cancer. I gave her massages a couple times a week. For months. My brother made a tape of her favorite songs to play while I worked with her. Laurie Anderson's album *Strange Angels* was on it."

I take a deep breath.

"The song came into my head as soon as I started pedaling after my husband dropped me off in the UP. It's always there, and if I concentrate it's like clicking on a tape player. I can sing along as I ride if I want. The lyrics just come out."

I can't stop.

"My dad passed last summer. It was a long battle and a dreadful death. I helped him die. I'm sure I'm working out my grief during this trip, hard not to do with pedaling alone all day."

Shane listens. Patient, he is simply with me.

STATS
27.32 miles, max speed 27.32, ride time 1:59, avg speed 13.6, TOTAL 1976.56

Synchronicity again. 1976. '56 is the year of my birth. Riding home is the right thing to do. Andy knows I am less excited about the celebration and more eager to turn my bike east. "You need to do it and have fun." I will try, I just want to fly.

Shane at the counter with my homemade pasties.

37

Until This Moment

Friday, July 15, 2016
East Missoula to Missoula, Montana

The girl researched and readied her gear. Backpacking tent with a rain fly, North Face sleeping bag, Svea cookstove, fuel bottles, panniers, handlebar bag, leather cycling shoes. Maps of each state on the B'76 tour ordered, bicycle tuned. When the snow melted, she rode to work, the fifteen miles to Stoney Creek Metro Park (with six-mile laps) and home, anywhere and everywhere.

Her mother took the girl shopping. "You should buy this red Adidas sweatsuit; it could come in handy on cold days."

Her father, reading the Detroit Free Press after dinner, said, "I don't know, sure seems like a lot of bother just to go on a bike ride."

The girl gasped.

"Sometimes the anticipation is as exciting as the event itself," her mother said.

Mikey was almost right. He said after two days in Missoula I'd know what to do. Not there yet and I already know.

"So, have you decided?" Mom asks when I FaceTime her early.

"I'm riding home."

She chokes back tears.

"I'll be back before you know it." First? Get through this weekend celebration.

Erika insists I stay for breakfast. "Start your laundry, and then I'll drive you across town to REI. The boys want to come too."

On my list: Coleman fuel and two tent stakes. Oh, and pants. Neither my faded tights nor oversize shorts will cut it, even if dinner isn't formal.

Missoula is a bustling city nestled at the northern tip of the Bitterroot

Valley, between the Bitterroot Mountains to the west and the Sapphire Range to the east. Buckled in Erika's front seat, hurtling along at seventy miles per hour on the expressway, I pant like a trapped coyote. Where's my helmet? Don't look out the side window, keep eyes straight ahead.

Erika glances over. "You ok?"

I relax my grip on the door handle. A little. "Not used to the speed I guess."

We make it. After picking up my supplies and a few extras, I choke on the $65 minimum price for a pair of women's pants.

"Hey, I'll bet you'll fit in a kids' size," Erika says.

Sure enough, kids' size eighteen is perfect. The lightweight brown pants that zip off into shorts are a tad long, but a bargain at twenty bucks. Inside, I'm still pudgy; outside, I guess I'm not. Yesterday, fully clothed, I weighed in at 123 pounds on Erika's scale. Before I pedal away on my final leg to Missoula, she feeds me a turkey sandwich.

A few miles down the road a sandwich board sign outside a Great Clips franchise advertises $10 haircuts. Time for it, my hair pokes through my helmet vent holes. The tang of permanent solution hits me with a memory when I open the door.

I sat on the clammy basement cement floor with the warm washing machine rumbling against my toddler back. A basket of fluffy toy kittens occupied me while the snip, snip, snip of scissors rained tufts of hair. My mother, a trained hairdresser, supplemented our family's income by styling the neighborhood ladies' hair. Her basement salon always smacked of permanent.

"You played so well by yourself," Mom liked to say when I was older.

She cut my hair too.

"Come on Mom, cut it shorter."

"No, it's short enough. I get so embarrassed. I tell the neighbors you are on a swim team, that's why your hair is so short."

"Why don't you tell them I have cancer, they'd have more sympathy for you." After B'76, I found someone else who would cut my hair short enough.

Inside Great Clips, a male stylist flips a black plastic cape over me like a matador. "What would you like today?"

I stand firm. "You can't cut it too short. I hate it in my face, I like it cut up around my ears, and I don't care what the back looks like or if it sticks up. I don't do anything with it, I just wash it and finger brush it straight back."

He gives me a perfect cut.

The ACA headquarters in downtown Missoula is easy to find. Bicycles are stacked at bike racks, against trees, cabled to parking meters. People wearing jerseys and bike hats line up on the sidewalk to pick up tickets for tonight's reception and tomorrow night's dinner.

"Tomodachi!" Towering over the crowd, a well-over-six-foot man with the B'76 logo tattooed on his calf waves me down. His broad, snowy beard obstructs the same symbol and the words "Bikecentennial" on the front of his original B'76 t-shirt.

"Big Dave!"

Big Dave was in our B'76 west-to-east camping group (harmonica always handy). An Ohio native, he now lives in Missouri. Before social media made it easier to reconnect, we ran into each other at a few bicycle tours in Michigan when Andy and I owned the bike store. In 2001, Dave traveled to Missoula for the ACA's twenty-fifth reunion. "I couldn't miss this one," he had messaged me on Facebook.

Tomodachi is the Japanese word for "friend." Yuichi, a Japanese rider who landed in Oregon instead of Virginia as he planned, dubbed our B'76 group the "Tomodachis" after we welcomed him into our midst. After dipping our front wheels into the Atlantic Ocean, Yuichi kept riding. When he rode through Chicago that fall we had a mini-reunion at a group member's house; that Christmas I flew out to California to visit the tandem team and Mike. Everyone watched my hour and a half film and no one laughed at me lugging a Super8 movie camera across the country.

"I haven't seen Rich yet, he said he'd be here at 2:00." Dave lifts me in a big hug which doubles as a method to plop me cuts in line. When

Rich found out Big Dave and I were coming to Missoula, he arranged a get-together. “I can’t stay for the parties, but we can have an early dinner Friday.”

In 1976, Rich alternated between riding as an independent and a group leader for those taking part in two-week sections of the TransAmerica Trail. His schedule coordinated with our group much of the way. Rich returned to his home in Spokane afterwards and worked as a writer and photographer for the *Spokesman-Review.* He is now its Outdoors Editor.

Touring cyclists are used to sharing spaces. When the Tomodachis gathered in Oregon, we divided group gear. I carried the first aid kit, someone else packed away the cook stove, another carried fuel, two others strapped five-gallon pots onto rear racks. Tents got shipped home when those who came alone paired off to share one. I didn’t hesitate to share a hotel room with Big Dave when he suggested it last winter. At some point I realized Andy might be uncomfortable with these arrangements, even if it saved money. How would I feel if he shared a room with another woman? I guess it wouldn’t matter (unless he was with one of his exes). In time to make alternative plans, I asked Andy if he as okay with it. He said he thought it was weird at first, but then admitted it was not a big deal.

And it isn’t. Big Dave is not an ex. Now Rich, he’s not an ex either, but I would not have agreed to share with him. Rich and I had fun in 1976. Not as much fun as my mom worried about, but I remember slow dancing with him in a bar during a rest day in Missoula…

Lanyarded with ID tags and party tickets, we take pictures until Rich finds us. “Let’s meet at this Mexican restaurant a few blocks over,” he suggests. *Fine with me, I’m starving.*

The three of us reminisce about B’76 and chat about being back in Missoula. The “bike tourist’s Mecca,” as Michael Prest called it ages ago when I sat with him at the million-dollar rest stop in the middle of Montana’s eastern prairie. Missoula, where the famous Sam Braxton, of Braxton Bike Shop, rebuilt my rear wheel forty years ago, and where I bought my first pair of cycling shorts (black wool Kuchariks lined with a leather chamois). I remember wondering why I had cycling shoes but waited so long for the comfort of padded shorts. I remember sweating with a hangover in a steamy laundromat.

"I remember the back-rub slow dance," Big Dave pipes in. *He does? Me and Rich? Or did I dance with him too?*

We speculate about our Tomodachis. Except for us, and a couple independent riders we met, online searches turned up no one.

"I saw Colleen and Shirley at the twenty-fifth anniversary," Big Dave says. "They moved to Missoula after the ride."

Shirley was a forty-something, divorced mother of what, six? Colleen, her youngest, was eleven. Shirley got excited, then disappointed, when we met in Eugene. She thought I was a lot younger and would play with Colleen.

How do you catch up on forty years over dinner?

Rich shares stories of a life outdoors and raising two daughters with his wife. Dave talks about his love of bicycle touring, years of being sober, winning a bout with cancer, and looking forward to retirement from a factory, where, as he describes it, he babysits gas boilers and compressors by computer with "babbling baboons (crazy co-workers)" in a "circus salt mine" (what he defined as "show time on the job"). Having worked in a few factories myself, I can identify with his creative coping skills.

Happy my old friends lived large since 1976, there is no residue "what if's" for me. Yet, I half expect Bunny to bite the back of my neck when I list my bicycling exploits after that summer. Blah, blah, blah, I've done this and I've done that. Bragging? Or just relating? It's me. It's not me. *Have I learned nothing?*

Rich springs for the meal of which I can't seem to get enough. I scarf the endless basket of tortilla chips. Until now, food was just fuel. Is my body recovering with these shorter mileage days and now begs for reloading? Or is my insatiability an irrational something else?

I am lost. The open road calls.

We take a selfie outside the restaurant, me in between the guys holding my original B'76 bicycle cap. I give Big Dave the flag I bought in Jamestown, Virginia at the end of our trip, signed by the Tomodachis. For Rich, an extra t-shirt from the bunch I had printed that fall. I copied what Yuichi wrote on our reflective triangles: Tomodachis in Japanese in front of a round sun, the English word and '76 beneath it. Mementoes held for forty years and carried all the way from Michigan.

"This part, at least, Bunny, I've got figured out."

Nostalgia charges the reception hall. More than 600 of us cheer through the original film made to promote B'76. Founders Dan and Liz Burden and Greg and June Siple open the presentations. I jot notes as fast as I can.

Dan refers to the "'76ers" as dreamers who carried out their dreams. He tells of the near impossible task of coordinating over 4100 riders coast to coast and shorter distances in between. I love when he quotes Karl Wallenda: "Being on the tight rope is living! Everything else is just waiting. Once flying, we no longer wait. Our talent, our essence comes alive." Substitute tight rope with bike and flying with biking—it's the same.

June calls us her "bicycling tribe." She echoes my own thoughts. "Bicycle touring teaches about geography, tolerance, brings us down to earth, whets our appetites, brings determination. It teaches us how to live and find treasure in the lives we lead. We can do anything!"

Greg talks about synchronicity. "What if I hadn't embraced the bicycle? You never know what the smallest encounter will bring."

Indeed. Bunny tickles my throat, I can't swallow.

Dave taps my arm. "Look around the room. There sure are a lot of old people."

I look up at him. He sees my damp eyes and grins behind his beard. "Dave, it was forty years ago, we're all old."

And we are '76ers.

Until this moment I didn't realize what that ride has meant to me—how it shaped my view of people; that when we strip BS away, we are all the same; how it grew my love of this large and wondrous country; and, not least of which, how much I love Michigan as my home.

I am a '76er. And proud of it.

Entertainment begins. I hover, watching myself outside of myself. Life is a long, wide gulf from there to here. I face the sunset now, no longer the girl I once was even if some essence of her lingers.

Forty-years ago, on the road over Lolo Pass the day our group cycled into Missoula, we met a young man walking across the country with a dog and a pony pulling a cart. He started on the east coast two years earlier. "I figure I'll be on the road another year," he said.

I daydreamed. *"If he asked me to leave my group and come with him, I would."*

He did not. I doubt he even noticed my existence. I thought of him often, especially when I returned home and couldn't sleep on my bed after three months on the ground. How could a person adjust to "regular" life after three years?

The open road tugs at me, whispering promises of a freer life, a life of going, taking the next turn to see around the bend, stopping here and there with no restrictions. Now I am torn. The road ahead is still open, still calling, and I am still eager to answer. But. This thing I have with Andy? It draws me to him like the moon draws tides.

I need to ride.

Home.

STATS
12.14 miles, max speed 24.5, ride time 1:28, avg speed 8.2, TOTAL 1988.71

Walking man, Lolo Pass in 1976.

38

My Tribe

Saturday, July 16, 2016
Missoula, Montana

Thrilled to begin this bicycling journey, the girl was eager and anxious to meet the group of strangers she'd be riding with for the summer. It was her first trip in an airplane, first time embarking on an adventure without someone she knew at her side.

As she unpacked and assembled her bike from the shipping box, she cast glances at the others. The tandem couple looked expert with their hand-sewn panniers and Bell helmets, the assistant leader professional in his black wool cycling shorts and jersey. Several others had helmets too, with small mirrors attached.

It surprised the girl to learn that of the fifteen riders in her west-to-east camping group, she had the most experience—a one-month tour around Michigan two years prior and two-weeks with her younger brother the following year.

This fact made her smile.

Tentacles tease from the back of my head and snake through my skull in a slow burrow forward. Flat on the hotel floor, my face smashed into my sleeping-bag-duffle-stuffed-with-clothes pillow, I know what's coming. A right eye tingle, a turn and my head will follow my face like a thumb flip through a thousand-page book, a knife thrust through my forehead. My twisted version of a migraine.

I ease myself up in search of ibuprofen, hoping to hold off the knife as long as possible. Sometimes these headaches linger for days. Interesting one shows up now.

Dave is up and ready to roll.

"Got time for breakfast?" I ask "My treat."

Dave has a bus to catch at Silver Park, site of a bicycle expo later this afternoon. He and other riders paid for a bus ride to locations south along the fifty-mile Bitterroot Trail. He'll ride north to a ribbon-cutting ceremony at the Travelers Rest State Park in Lolo. I'll join him there for a group ride on the last section of the trail into Missoula.

"Sure, let's grab something here."

We repeat last night's clown act in the elevator. Dave squeezes his rented mountain bike in and takes it down four floors, returning for a recumbent assist. Good thing he's tall. He lifts the front end of my rig into the back corner of the elevator, I shuffle in the rest. It barely fits the diagonal from floor to ceiling.

What I am not used to: traffic stopping for a bicycle at every road crossing, intersection or not. I can't help it, I stop at every crossing, and every time an oncoming car stops too, they wave me through. With a smile, I might add. *Am I still in the United States?* This would never happen in the Motor City, or anywhere else I've ridden. Well, maybe France. During PBP, if Lou and I stopped for any reason, passing drivers pulled over to help.

Not far out of Missoula proper, the bike trail, bordered by fresh dirt that needs seeding, is pristine asphalt. I cruise a green valley surrounded by green mountains. Far below, the Bitterroot River flows along a railroad track, Highway 93 races on the right. A striking view, the smooth and relaxing ride holds migraine at bay.

Traveler's Rest State Park is, according to the Montana State Park website, "The only archeologically verified campsite of the Lewis and Clark Expedition in the nation." The verifiable evidence is remnants of their trench latrine. I leave archaeological evidence almost everywhere I go, too, but doubt anyone will open a park for me.

I wander as riders and other tourists gather for the ribbon cutting. No fruit or power bars is a disappointment, especially since I left my food bag at the hotel. Lucky there is plenty of free water. My head is light with no escape from the blazing sun.

At last the ceremony begins, but speech after speech drags on. An announcer finally, finally, calls the contingent forward. "'76ers, raise your hands!" Big Dave and I, and at least a couple dozen others, raise our hands. "Riders, please let the '76ers lead the way!"

Like the Red Sea, the congregation parts. We '76ers roll under an arch of red, white, and blue balloons, followed by hundreds of cheering cyclists. Emotion chokes me by surprise. I pump a fist at a barrage of clicking cameras. What I remember in 1976—being part of something larger than just a bike ride—is rekindled. Joy. Love. Fulfillment. The "I can do anything" confidence, the brotherhood of the road, the "we are all in this together" sense of things. June Siple's words echo. These people are my tribe. How could I have thought this celebration wasn't a big deal? I am thrilled I rode here.

Silver Park is full of bikes lying on grassy knolls, leaning against trees, walked along by their riders. Spaced along the trail are vendors of all sorts from REI to the Nature Conservancy hawking free swag. I snag a new water bottle.

"Patti! Is that you?" The gray-haired man walking toward us looks familiar, but I can't place him. "It's Mike! Mike DiGregorio!" A grin lights his face.

"Mike!" We hug. The DiGregorio family lived near my childhood home in northwest Detroit. Mike is a couple years older than me; his brother Frank was in my grade at Immaculate Heart of Mary Elementary School. The DiGregorios later moved to Sterling Heights, where my family ended up for my last year of high school. Mike loved bicycling too, but we never hung out together. He left on a B'76 camping group a few days before me.

"I flew out...for the anniversary. I came for the thirtieth one too. I've been walking. All day. My hotel…is on the east side of Missoula. About six miles away."

Mike's stumbling speech is peculiar. I tell him I rode from home, but he pays no attention, he keeps talking. And smiling.

Big Dave backs to leave. "See you at the hotel, Patti."

Mike enumerates the last forty years of his life. His wife died of cancer "some years ago." He has kids. "Two grandkids are my life." He lifts his shirt to show me a Walkman-sized gizmo on his belt, with a wire leading to a patch on his belly. "This…my pump. I'm diabetic." Insulin pumps deliver insulin automatically into the body at a programmed rate, or manually when more is needed.

In 1977, we both applied for work with Ma Bell. Mike took an office job. By the time I got called for a linesman job, I was making a dollar an hour more at the aerospace manufacturing firm and stayed put.

"I retired from Ma Bell. Had a brain tumor removed. A little trouble with memory…and finding words."

His stop-and-go speech makes sense to me now.

Active in his church, Mike lives in the house where he raised his kids. "My daughter is always…worried. About me. But I'm grateful… for every day I have."

Happiness bleeds from this man.

"Can I take a photo of us to send my mom?" I ask. "She'd like to hear I ran into you."

For a moment, Mike is interested to hear about my folks. His have passed. With the click of my phone the knife stabs above my eye. Hot sun and bonking prove too much for a couple ibuprofen.

"Gotta get going. Was great seeing you, Mike."

I wrangle my bike into the elevator myself. Big Dave is showered. "I'm going down, meet you at dinner."

Imploding my head is. Guzzle water, more ibuprofen, take a cool shower. Ready or not, I slip my new "boys" pants (this thin and wiry body doesn't feel like me) and head out, glad I don't have to go anywhere else.

Big Dave is easy to find, towering among the crowd of at least 500 people. I grab a seat and empty the filled water glass at my place, happy to see several full pitchers on the white-shrouded table—and the army of waiters staging stainless steel covered bins onto mile-long buffet tables.

"Hi, I'm Bryan, this is my wife, Patty." The couple next to me are from Seattle, Washington. They aren't '76ers, but they've done their share of bicycle touring with the ACA.

"I think I'll remember your name." We shake hands and I slug another glass of water.

"Welcome to Adventure Cycling's 40th Anniversary." From a stage beyond the dinner tables, a man waves his arm and bows. "First things first. Please, help yourself to our feast."

The waiters lift lids from piles of taco shells, tortillas, rafts of refried beans, shredded beef, chicken, and all the fixings of a Mexican buffet. No one moves. The announcer leaves the stage, he's not calling tables.

"Heck with this, I'll start!" I don't care I ate Mexican yesterday. First in line, I snatch a taco shell and two Frisbee-sized flour tortillas. To expedite the tsunami that breaks behind me, I pile ingredients into a mountain on my plate, leaving construction for back at the table. Ignoring sideways glances from Patty and Bryan, I excavate my way through everything. No need for seconds. I lean back, migraine calmed to a backstage throb.

Tonight's program focuses on ACA achievements since 1976 and continued expansion of bicycling routes. Featured speakers are Lael Wilcox, winner of both the Great Divide and TransAm self-supported races (probably a good thing the ACA didn't offer these events in my ultra-marathon days); Erick Cedeño, "The Bicycle Nomad" and founder of the Bicycle Nomad Café in Phoenix, Arizona; and Willie Weir, a long-time writer for the Adventure Cyclist Magazine.

Lael's and Erick's words echo my lifelong love affair with the bicycle: an elegant blend of human and machine, a vehicle of discovery and freedom. Willie speaks truth about finding commonality on the road. I reflect on comments to my Facebook postcards. With the disheartening political climate ramping up to this fall's election, the kindnesses I encounter give hope. By connecting we overcome divisiveness. Maybe more people should ride bicycles.

To close the program, our voices rise in bicycle-themed lyrics to a re-written "This Land is Our Land." Look! As if on cue, a full rainbow appears across the cloud-dotted sky.

STATS
28.71 miles, max speed 28, ride time 2:58, avg speed 9.6 TOTAL WEST 2017.47

My 1976 Bikecentennial ID.

39

Extended Reunion and a Kindred Spirit

Sunday, July 17, 2016
Missoula to Ovando, Montana, heading east now.

Some older workers at her first factory job disappointed the girl, surprised to realize that age did not always bring wisdom.

"Seems like once an ass-hole, always an ass-hole." A cynic at age twenty.

And yet. Here, an entry from her journal dated May 30, 1976:

We covered twenty-two wet miles and never found a campground—Brad (eleven) found us and kidnapped us to his house—we stayed the night—fantastic people! Blueberry pancakes for breakfast and everything. GREAT! If people in the rest of the country are anything like those here—wow.

Street barricades portend mobs for today's BIKEapalooza, third-day activities of the bicycling celebration. The migraine knife-point balances against my bloated forehead. Will the open road and fresh air chase the decay? I need to break away. Like a caged dog pacing, I plot my escape.

Dave is staying another day. "I'll ride you out of town. But first, breakfast on me. And not here." One of his friends pointed him to a downtown cafe tucked in a historic building with a small casino. Its stressed plank floors, dark paneling, and low lighting are perfect.

"Thanks, Dave," I say. "And not just for breakfast." I thank him for splitting his room, for sharing my story. *He's a great example of how bicycling can expand a life.*

Before the east side of Missoula, Big Dave lets me go. The rising sun reflects my smile. *Eastward ho!*

I swing by Erika's house and hang a bag on her doorknob with gifts for her boys—the metal cup from yesterday's ribbon cutting, an ACA 40th Anniversary bicycle bell—and Simon's bike lock. He lent it to me after the lightweight, resettable combination lock I bought at REI wouldn't open. Someone must have changed the factory setting. All these miles and days with no trouble, but warnings about leaving bikes unattended in Missoula worried me. Not sure I could bring my bike inside the hotel, I didn't want to take the train home because my ride got stolen. Unfounded concerns, I never used Simon's lock.

"Thanks for everything." I roll a whisper down Erika's drive.

The sky is clear, air chill enough for fleece. The road winds up an easy grade between hills along the still-sparkling Blackfoot River, and yes, an east (but gentle) wind. Headwind Queen's personal Murphy's Law of biking: ride five weeks west with mostly headwinds, turn around one day and the wind shifts east.

Ah, well. Nothing to do but empty the mind and pedal.

Here is the road that lifts to Sundog Permastead. I give a nod to Julie and Hunter. Tonight, Ovando. Tomorrow I divert from retracing my westward path to stay with the Newell's in Helena. From there, south to Three Forks to pick up section eight of the Lewis & Clark ACA route and follow the Yellowstone River back to Glendive. North Dakota a repeat, and Minnesota with its lovely bike paths. I'll map my own course into Wisconsin to visit Karen, then east to Manitowoc and the ferry to Michigan.

A new goal for this ride—getting home by the weekend of August 13-14, when Mandy and Josh visit—reflects how my bicycling life morphed from a carefree-Forrest Gump-pedal to a competitive, "put in the miles." Getting to Missoula on schedule was a confidence builder, and this old body, adapted to riding daily, is eager to hammer. No more twenty-five measly miles in a day. Despite this east wind, I hope for a westerly push, ever mindful of the altitude loss each mile toward home.

My phone rings just as I pull into the rest stop where I met a grieving Joanne and her dog, Lolo.

"Hey Patti, It's Dave. I found Colleen! Here she is."

"I'm so sorry I missed you," she says. "Where are you going to be tonight?" Colleen knows where Ovando is. "I want to see you. I have things to do this afternoon, but Big Dave and I can drive out around dinnertime."

Thirteen miles to Ovando fly by. Two young men, traveling the Great

Divide Trail independent of each other, arrive at the same time. I claim my spot in the hoosegow. They opt for the tepee and shepherd's wagon across the square. Lots of time to settle in before Colleen and Dave arrive.

Big Dave extracts his frame from Colleen's mid-sized sedan like a hippie Transformer, colorful in his tie-dyed t-shirt and matching ball cap. Colleen is a lifetime of change on the spot. The last time I saw her she was pre-teen cantankerousness topped with a wild mop of sun-bleached hair, our nine-year age gap as vast as the Grand Canyon. Before me stands a fifty-one-year-old natural woman, wife, and mom dressed in khaki shorts and sandals, a loose-fit printed blouse puckered low around her neck, a woman I suspect could be a good friend.

I instantly regret my twenty-year-old aloofness.

"I brought my old photos." Colleen hefts an album barely holding together. We three sit at the octagon picnic table in front of the hoosegow, turning every page. Faded color photos spill memories. "I whined the whole way," Collen says. "That first day, it was only seven miles. Remember we dipped our rear wheels into the Pacific? I told my mom there was no way I could ride fifty miles. It wasn't until my thirties that I realized what she had done."

Shirley, gone now almost ten years, dragged her daughter across the country with only a twelve-mile training ride under their belts (on brand new bikes) only ten days before our group left Eugene.

My stomach growls. The Stray Bullet Cafe is long closed. Our only choice for dinner is Trixie's Bar at the edge of town. We pile into Colleen's car and find a full parking lot. *Where did everyone come from?*

A "Welcome to Trixie's" sign hangs by chains from a natural log perched high on two painted logs at the entryway. The saloon is part bar, part restaurant, and part casino. Mikey's observation again, casinos everywhere. A waitress seats us in a quiet side room.

"Get whatever you want," Colleen says. "It's on me." I choose protein in the form of a rare steak. The tinge of guilt I feel when Colleen orders a hamburger is extinguished with Big Dave's steak and lobster combination.

Stories flow. Our conversation moves to our respective forty years of living. I sense a sadness as Colleen talks about her home life. Her youngest son is leaving for college. *Empty nest?*

Guilt returns when the bill comes and Colleen excuses herself to withdraw money from an ATM by the door. "I didn't bring enough cash."

Ugh. I offer to help pay, but she brushes me off. "No, it's fine." I am not so sure.

Back at the hoosegow, a balding man with glasses, wearing a neon-yellow bike jersey over baggy shorts, reads at the picnic table. A one-person tent is set up on the grass under a tree. Well, it looks more like a bivvy sack than a tent. He travels light.

"I met him in Missoula." Big Dave unbends himself from the car. "He's from Michigan and has written books about bike touring."

The man points his book at a graying black lab standing atop the table. "Meet the Mayor of Ovando's dog." The dog's tail whips in circles like a cowboy's lasso.

"That dog held his own leash when I saw him here a few days ago," I say. The dog walked next to a middle-aged man dressed in dusty jeans and a ragged shirt. "You must have talked to Kathy at the outfitters. She said the 'mayor' is special needs and the whole town looks out for him."

We trade introductions. Bob Downes started in Seattle and is riding home to Traverse City, Michigan. "There are two bunks in the hoosegow, do you want to stay inside?" I offer.

"No, I'm okay out here." He graciously takes a photo of our Tomodachi threesome before Big Dave and Colleen head back to Missoula.

When storm clouds darken the evening, Bob crawls into his mini-tent. I head over for a last stop at the porta-potty tucked behind the shepherd's wagon. The younger of the two off-road riders perches on the wagon's top step, writing in a notebook. Jimmy also hails from Traverse City, but he's been living on the west coast a while.

I linger to share a moment of awe watching lightning dance over the mountains. "We were in Colorado in '76 on the Fourth of July when a storm like this blew in. I'll never forget it, made the fireworks look lame."

We chat about traveling, of adventure on the open road.

"I've given in to it, I've stopped fighting it." Jimmy works whatever job he can find, often in a bike shop, to save enough money for another trip. "I'm planning seven months for this one. When I reach Mexico, I'll keep going. My parents don't understand."

I do.

The storm stirs me. I share my Forrest Gump ride-until-I'm-done story, my mixed emotions and indecision about riding home, how glad I am to be riding back. "Do you know the exact moment when I made up my mind?" I tell him of the black wall, ducking under my fairing to lessen the sting of rain and hail. "It was right then, right when any normal person would've decided to stop riding, right then I decided I must ride home."

Jimmy nods his head, understanding completely the foolishness of loving the most uncomfortable and perhaps the most dangerous situation of my ride. "Do you think you'll be done when you get home?"

Kindred souls, we stand spirited beside each other in the magnificent darkness. I kick a stone and gaze a long moment at the mountain light show. After "three years, two months, fourteen days, and sixteen hours" of running, Forrest Gump stopped. "I'm pretty tired," he said. "I think I'll go home now." I am anything but tired. I am motivated to ride. Home. Yet, this nomad life.

"My ride isn't what I imagined, but I'm okay with what it is. I think it will be enough."

He faces forward. "Nah, you aren't done yet."

I wonder if he sees my smile.

HEADING EAST! STATS

53.75 miles, max speed 34.5, ride time 4:36, avg speed 11.6, TOTAL 2071.30

Big Dave towers over Brad and his parents in Oregon. Our group leader Steve is in the background.

40

Getting my Groove Back

Monday, July 18, 2016
Ovando to Helena, Montana

The girl floated, somewhere high.

Peace.

Her gaze turned. Was that her bloody body below, slumped between two men on the bench seat of a pickup truck?

She fought to remember.

A flying descent on a Kentucky mountain road, sweeping curve, pavement gone. Loaded touring bike launched off a six-inch break, landed upright in gravel, and kept rolling. Until pavement reappeared. "Oh, shit." Attempted wheelie. The handlebar bag broke free and tangled in the front wheel. Last thought: "Get away from the bike."

The girl watched the man-not-driving lift a scratched Super 8 movie camera from the bag. "The lens is loose, but it still opens and closes."

"There's film in there!" She shrieked, but no sound escaped her mouth. No longer floating, she refused to open her eyes.

No surprise, another headwind. Poor Bob's bad luck, running into the Headwind Queen; he's hoping for prevailing westerlies to spur his straight hammer back to Michigan.

At SR 141, I divert south, a gentle rolling cruise along Nevada Creek. The Garnet Range rears west of the river, Helena National Forest stretches east. No driveways, no fences, no people. Nothing but low pines and wispy clouds above the peaks in a surf of blue sky.

Foothills rise. Foothills fall. Rise and fall are contradictory with effort. What looks uphill should be harder, yet demands a higher gear. Where I expect a downhill assist, my speed slows. I recall similar disorientation with Lou somewhere in the southwestern desert during our 1988 crossing.

Back in my crazier days, I fantasized riding RAAM, a non-stop, fast-as-you-can (with little sleep) jaunt from the west coast to the east. To test my competitive ability (given my asthma), Lou and I signed up for a PAC Tour "training" ride. PAC Tour, offering fast, supported transcontinental bicycle tours, is the business of RAAM veterans Lon Haldeman and Susan Notorangelo.

We rode in mid-August, the hottest month of the hottest summer of the decade. In the desert, no escaping the sun, we draped socks full of ice cubes around our necks. On one stretch, we floundered. "Why is it so hard, Lou? It looks like we're going downhill, doesn't it look downhill?"

I wheeled a u-turn. In one pedal stroke, the tandem surged and we flew past another rider from our group fighting the grade. "Where are you two going?" he hacked.

"Not this way!" I hit the brakes.

The heat took its toll. We weren't even out of California when I decided RAMM wasn't for me. Too much effort and expense just to say you did it. But we kept on with PAC Tour. We rode, ate, rode, ate, stopped, checked the bike, slept, got up, and did it all over again until we dipped our front wheel into the Atlantic Ocean.

"So, Bunny, another lesson? Forget what I see, go with what I feel?"

Now what feels like a climb looks like a climb. Nevada Creek spills into Nevada Lake, or perhaps it's the other way around. The road winds high above the miles-long body of water, hills spew the smell of pine and spread toward blue-tinged mountain ranges in the distance. Where the lake narrows, a couple dozen vulture-sized birds peck at the edge of a sandbar. White on blue, a reflection of the sky. One or two stretch wings, exposing black end feathers. Pelicans surprised me in North Dakota, are they here too?

"Might make for an interesting backdrop." I pause at a guardrail and break out my Nikon. Self-portrait with timer. *Who is that thin woman squinting from the screen on the back of my camera?* Taut thighs, puckered brow, she clinches her bicycle helmet under her arm as if for dear life.

SR 141 drops into Avon, an almost deserted town, dusty on the banks of the Little Blackfoot River. At the turn onto US 12, before tackling the 6320 feet of MacDonald Pass, I make a quick decision to treat myself at the Avon Family Cafe.

The cafe is a vertical log building with a steel roof. Large windows overlooking the gravel parking lot are inviting, even if the view of the highway and railroad tracks is not. A few tables, fewer patrons. A long counter fronts an opening to the kitchen, above which hangs a wooden sign, "Welcome Friends." The place must also serve as the town's bakery—a glass case is filled with iced cinnamon rolls (bigger than bricks) and an assortment of pies, cakes, and cookies towers opposite the register. I don't know why I resist.

The lone waitress is a gray-haired woman with a flour-dusted purple tunic for an apron. She brings me a one-page paper menu printed on both sides. "Coffee?" She gestures with a full pot.

"No, I'll just take water. Can I get my bottles filled too?"

"Sure," she smiles.

My BLT comes on homemade wheat toast and the cook does a great job incinerating the bacon. I gorge but dawdle, maybe a bit nervous about MacDonald Pass coming at the end of a long-mile day. The top is fifteen miles or so from here, I have no idea where the climb starts.

No more excuses. Time to go. Right away the road rises, mild at first. After eight miles, two lanes split into four and the real climb begins. I alternate between pushing 100 steps, resting, pushing 100 more, and pedaling the switchbacks. After 2000 steps counted, I lose track. Just before crossing the Continental Divide, a dirt road exits right, crosses a cattle guard, and continues up into the looming mountainside. Cromwell Dixon State Park Campground. Bruce said it could be a decent stop, "But it's primitive, and exposed if the weather turns rough." Funny he mentioned that. Gone is the open blue sky with wispy clouds. Gray billows threaten.

At an overlook before the descent, I dig out my rain jacket and turn on my blinking red taillight. A car pulls in right behind me. A couple, puffing on cigarettes, gets out to view the vista. We are above everything, the undulating valley, peaks blanketed in pine forests.

"Can I bother you to take my picture?" I ask the woman when she inquires about my ride. She snaps a few before we feel sharp stings of rain and she races back to her car. I hurry to secure my camera and take off conservatively, pumping brakes to keep my speed below twenty-four miles per hour. Running full out on fresh wet pavement is risky, even if this weighted, long-wheelbase recumbent is super stable.

Am I outrunning the rain? My hands relax, I lean back to give in to gravity. Rolling down the mountain, my singing of Jethro Tull's "Fat Man" pushes Laurie Anderson to the background. It is a glorious, luge-like run into Helena.

Bruce wasn't kidding about the last mile to his house. The steep hill forces me to dismount and push. A block away, the sky lets loose. Ah, well.

The Newell's house sits at the curve of a cul-de-sac in a hillside suburban subdivision. Susan opens the garage door for my dripping load. "Bruce is out on a nine-day ride with some friends, but I'm here. Come on in! We really don't have a spot for you to set up your tent in the yard. You're welcome to the deck, but we have a guest room downstairs ready. If you stay on the deck, you might have to fight the bunnies for space."

Rocky ground slopes from their wrap-around deck. And yes, there are bunnies. Tempted to hang with them, the cozy guest room (with a bathroom right around the corner) beckons. *Sorry Bunny*. Inside it is.

Susan knows how to make a fellow cyclist feel welcome. She rustles up a filling mix of salad, a casserole of bread, cheese, and mushrooms, and homemade banana bread for dessert.

"Care for a glass of wine?"

"Sure, why not?" First wine of the trip.

We spend a relaxing evening sharing tales from the road. "I'm so happy to 'pay it forward.' When Bruce and I rode across the country, and in Europe, people were so kind taking us in. It's nice to do the same for you."

Kindred spirits have a way of finding each other, I think.

POSTCARD FROM THE ROAD 7/18/16

The easy east wind is not so hard in my face and ears that I can't hear the squeak in my right shoe with every pedal push. For the first twenty-five miles.

The Garnet Mountains deceive. What looks down feels against gravity; what looks up feels like a sled ride picking up speed.

Jellyfish clouds with long wiggly wisps bob about the peaks. A bald eagle soars low ahead of me, following the road. He swoops to the shoulder and disappears. Glancing left, an antelope watches, its head turns slowly as I inch by.

Getting my groove back.

STATS

73.10 miles, max speed 36, ride time 6:47, avg speed 10.7, TOTAL 2144.43

Me and Lou on PAC Tour, 1988.

41

Delirious Murmurations

Tuesday, July 19, 2016
Helena to Three Forks, Montana

On the fifth day, an arduous, woozy day, the girl got back on her bike, tired of begging rides after her crash in Hazard, Kentucky. The tandem couple dedicated dinner to her. "You get the Heroic Biker award!"

The girl's smile cracked her hardening scabs. Time to call and break the news to her mother, now that she was fine.

"Patti! What happened? I sat straight up in bed last night, I just knew something was wrong."

"I'll ride out with you this morning, to show you an easy way through Helena." Susan serves a promise with fresh-brewed coffee and a heaping bowl of homemade granola.

We set off looking like Mutt & Jeff twins in our similar-color jerseys and white bike helmets. Perched upright on her hybrid, Susan towers over my low-profile steed. She leads a winding way through adjoining subdivisions and downtown proper. "Bruce calls this our 'sneak route.'" She takes a shortcut across a school parking lot to avoid busy intersections.

Helena is hilly.

"I'm glad you didn't take me the challenging way," I tease when Susan deposits me, seven miles later, onto an access road next to a highway south.

She laughs. "There ya go, on your way to Three Forks."

We hug goodbye and I head into the sunrise.

SR 287 follows a wide valley of green irrigated fields burrowed in with Canyon Ferry Lake between the Elkhorn and Big Horn Mountains. Grades are long and easy, traffic sometimes heavy, most times light, and surprise—the wind doesn't seem a factor, head or tail.

A few clouds bunch over distant peaks. Is that snow glistening there? A shifting glint catches my eye. A murmuration. Such a lovely word for the syncopated flight of birds. Sunlight reflects off the flyers' white feathers to create a massive undulating disk; a sudden twist and the swarm vanishes against the blue sky. Back and forth, white, nothing, then white again, the cluster breaks into three appearing, disappearing bunches. I crane my neck. There. A white V-formation, a dark V. And then, just sky.

The birds' on-again-off-again lift Mom to mind. When I called from the Newell's she said, "How can you impose on strangers like that?" Her nip was less about her not understanding my propensity to collect new friends, and more about her displeasure I'm riding home instead of taking the train. Nothing I say will make a difference. I am coming home after all. *What do you want, I'm not riding the long way home!* Gads, that's her voice coming out of me. Ah, well, let it go.

At tiny town Toston, I heed Susan's advice and take a side-road off the highway. "The road narrows right there as it crosses the Missouri River," she warned.

Sleepy. That's what this town is. Three suntanned men lean against a rusty pickup in front of a weathered house. Further on, a stand-alone Post Office. A girl and two young boys, one on a BMX bike, hang out on a slice of grass under a handful of trees. "Cool bike!" they chime.

"Thanks!"

Across the road is a gravel expanse before a triple set of railroad tracks. About 100 yards away stands a porta-potty. I passed one coming into Toston, but the door was locked. "Hey, do you know if that porta-potty is open?"

"We'll go check," the girl says.

Before I can tell them don't bother, I'll check when I leave, the boy on the bike kicks up dust with the other two in fast pursuit. His bike clatters on the gravel when he leaps off to rip open the door. "Yep, it's open!"

The kids disappear faster than the murmurating birds. They must not be like I was at their age, desperate to hear stories from a traveling bicyclist.

The Post Office is closed, no open lobby in which to steal a charge for my phone. With a wadded up raincoat for a seat pad, I lean against the brick building to jot a few notes in my "while-riding" notebook. The first bite of the Braeburn apple I picked up somewhere snaps juice over my words. Never heard of a Braeburn before Minnesota. Now it's my new favorite: crisper than a Macintosh and just as sweet.

Back on SR 287, a yellow road sign depicts a truck descending a 5% grade. The bike rockets. *What are those black things dotting the hills, a herd of cows?* Nope, looks like horses. *Wait, what? They aren't moving.*

If it weren't for the delightful pull of gravity I would stop. The horses are full-sized steel sculptures, arranged on the hillside in the middle of nowhere, as natural as could be.

The heat. It must be the heat.

What goes down must go up. At the top of a long slog to the intersection with Interstate 90 before Three Forks, the aroma of freshly baked bread turns me Pavlovian. Sharing a parking lot with the Fort Three Forks Motel and RV Park is the Wheat Montana Farms and Bakery. Perfect stop for the night.

A woman about my age comes out of the motel to greet me. "Our campground is out back, but it's really for RVs and not very nice for tenting. You might do better at the KOA." She reads my resignation. "It's a beautiful park. Lots of trees and shade and grass. You'll be more comfortable there." She either realizes I might be overheating, or she's bored. "Do you want some fresh water? Come on in."

The air conditioning hits me like jumping naked into a snowdrift after sweating it out in a stone sauna. If I stay here, I could hang out in the lobby this evening. After visiting the bakery. The woman hands me two icy bottles of water. She has a captive audience for her story of how a retired eastern city girl ended up in the frontier, "I was looking for adventure…"

Not much of her protracted soliloquy registers. One bottle empty, my stomach nudges for food. *The bakery*!

"Thanks again for the water. I guess I will head to the KOA."

Like a toddler overwhelmed at a penny-candy counter, I am mesmerized by loaves of crusty bread, muffins, cookies, chocolate covered brownies, croissants, turnovers, donuts, cinnamon rolls, and a wall menu listing homemade sandwiches. Thankfully the line ahead allows me time to regain my senses, one of everything is not an option.

"Can I get one of those steak roll sandwiches without anything?" No rolls in the case, one would go great with my Chicken Alfredo Compleat tonight. The salesgirl doesn't blink. "And a cinnamon roll." Tomorrow's breakfast.

Luckily, the ride to Camp Three Forks, "Where the Montana Mountains Meet the Missouri," is mostly downhill. Boiling, my brain still is. The couple who registers me could double as Mr. And Mrs. Klaus.

"Follow me." The bulging-bellied, white-bearded Mr. Klaus leads me out of the KOA office that, with its round tables and chairs, hand-dipped ice cream, couches and a big screen television, also serves as a gathering place and mini-store. He hoists himself into a golf cart. A graying hound clambers up next to him. Tonight's home is a soft patch of grass with a picnic table near a playground, complete with a freshwater spigot.

"If you need anything, just let us know," Mr. Klaus says. I half expect him to flick his nose and disappear. Instead, a black bunny hops by, trailed by three more. "A few summers ago some campers left their pet rabbit. It was black. I guess it found a mate, now we have them all over the place."

"Hey bunnies!"

STATS
70.52 miles, max speed 33, ride time 5:50, avg speed 12, TOTAL 2214.96

42

Hard Side of the Tracks

Wednesday, July 20, 2016
Three Forks to Livingston, Montana

Afterwards, the mother confided to the girl, "I worried you'd lose your virginity on that ride."

Not that the girl didn't have opportunities. Or desires. The scheme of marriage and babies kept her knees locked.

Billy Joel's song, "Only the Good Die Young" came out less than a year later.

"I'm just getting out of the shower, can I call you back?" Mom's voice is as sharp as the cloud bank that carves a narrow band of blazing sunrise above the Bridger and Gallatin Ranges, some forty miles away. Packed and ready to go, I set my butt on the picnic table, rest forearms on my knees.

"Hey bunnies, you sleeping in today?" Like with the wind, I made peace with Mom's rancor long ago. Still.

Her call comes, the edge in her voice dulled, "Before, I waited to do anything until you called, but I'm not doing that anymore."

Tired of worrying? Or fed up that I'm riding home? I close my eyes to remember the email she sent two weeks before this ride. An exhausting overnight shakedown, with headwinds (no surprise), a heavier-than-expected load, and an epic dinner fail (forgot to add yeast to my pizza dough and then burned it), shook my confidence. Mom wasn't happy I decided to ride my bike across the country again. Her unexpected words were a comforting surprise. "I still think you're crazy, especially for going alone. But it's your decision and I'll love you either way."

Avoiding the zig-zag ACA route, I hop onto Interstate 90's gradual grades to head straight for tonight's destination: Livingston, on the far side of the 5760-foot Bozeman Pass. Clouds break, temperature and traffic rise. Eleven miles west of the pass I stop at a gas station for a Gatorade break and map check. The ACA route, entering I-90 here until exiting before the summit onto a frontage road, assures, "You'll have a gentle climb over Bozeman Pass to get into the Yellowstone River drainage." I am skeptical.

A fellow touring cyclist rolls up and stands over his bike to chat. The young man is riding from Toronto to Vancouver. "The frontage road is good," he says. "And the climb is easy."

Bozeman Pass is a giant's three-step entryway, three obliging tiers of elevation gain. A drawn-out curve, a narrowing shoulder. (Did those mountains sneak closer while I wasn't watching?) Another curve and the sun sprinkles slim between Gallatin's rocky shoulders and Bridger's tree-studded slopes. *How is this ribbon of road going to slice between these looming peaks?*

Exit 319 to the frontage road a few miles from the top offers an escape from noxious, laboring trucks. The Toronto cyclist and ACA map are right, the climb (thus far) is kind. Whoops! The last half mile is so steep I hope my front wheel stays on the ground. Ah, well, press, press, press at fewer than five miles per hour. At the crest, the door opens and the road drops to rolling prairies covered in sagebrush. My reward for zero steps? A downhill run, right into Livingston.

Can that be a tailwind too? *Yeehaw!*

A thin man, dressed in green jeans, matching ripstop shirt, and a felt hat with earflaps tied at the top, exits the gas station convenience store where I've stopped. "Hey, where ya riding?"

"Home to Michigan."

He folds his arms to listen to my Missoula anniversary spiel. "I wanted to do Bikecentennial'76, but I chickened out." His cobalt eyes twinkle. "When I ride downhills, I hear Chopin. The faster I go, the faster the beat on the keys."

Before I can tell him about "Ramon," a woman marches into our conversation. "I picked him up at the Bozeman airport this morning. He made me take him to every bike store in town. And now here he is talking to you instead of getting going."

What is her accent?

"This is my wife." The man turns. "Hon, this gal has ridden her bike to Missoula from Michigan and is on her way home."

"I've been here a week with my mother at an art ranch up in the mountains," she says. "Bill's joining us for a concert."

Bill examines my bike. "I knew Gardner. We raced together way back when. I didn't like it though; he was so competitive and always beat me." Gardner Martin founded the Easy Racer bicycle company, the manufacturer of my Tour Easy.

I learn Bill is a county supervisor in California, instrumental in getting a law passed in his state to protect bicyclists. "Cars have to pass riders at least three feet away. We've had eighty fewer deaths since we passed the law. That's how dangerous it is out there."

His wife engages, "Have you ever ridden in Amsterdam? They ride bikes everywhere there."

Maybe that's her accent. "No, but I've raced in France. Same thing. I loved it and felt right at home."

The woman pinches Bill's shirt sleeve to pull him toward her car. He winks, handing off a business card. I bet he'd rather be pedaling to the ranch instead of driving.

The Livingston Campground and RV Park is on the hard side of the railroad tracks in this town of 7000. The ground is grassy and flat, mountains dominate the horizon. A ramshackle office is locked. Two men sweat digging fence holes to separate daily rental sites from a row of longer-term mobile homes.

"Need a site? You can set up anywhere over there." The younger of the two points to a narrow strip next to a privacy fence. "My mom will be back soon to get you registered."

There are no discernable site boundaries. A tent three times larger than mine stands at the far left, covered by a tarp tied to the branches of three scraggly trees. Nearby is a warped picnic table, a pile of firewood, and a department store mountain bike leaning on its kickstand. To the right, a smaller tent is tucked against the peeling backside of a garage. Except for the post hole diggers, the place is deserted.

I set up between the two tents and retreat into my own early to read, write, and relax. My resting heart rate is up to seventy. Even so, I hanker to hammer out of this huge state. Might depend on the heat, I've never done well with it. And my splendid old friend, the wind.

"Hey!" Rest interrupted by a gruff male voice, ravaged by smoking my guess. "You steal my table?"

Zip, zip. I pop my head out. A short, heavy-set man, who reminds me of an older Van Morrison, grips a brown paper bag. Greasy hair sticks out above his shoulders. He's missing an upper front tooth.

"Sorry. Didn't know it was yours. I'll move it back."

"No big deal. It's not really my table, I'm just renting it." He swings his leg over the seat. "I'm going to sit here and drink my beer and enjoy the night."

Should I dive back into my privacy or engage the guy? Maybe sharing small talk will make me a person in his eyes instead of a stranger ignoring him. No decision necessary, he gets chatty. He points to a Schwinn ten-speed that looks like it wasn't new years ago.

"I paid fifty bucks for it. I gotta lower that seat." He takes a swig. "I gotta get a new seat."

I can't imagine how he even straddles the bike, let alone sit on the seat to pedal, it looks tall enough for Big Dave. And, by the way, the seat is already down to the frame.

Another swig. "I was thinking of riding to Yellowstone, but I heard it was twenty bucks a night."

Until I met that westbound rider earlier today, I didn't realize Yellowstone Park is only a day's ride south from Livingston. Well, half a day for me. My guess is Van here, would struggle to ride even a couple of miles.

"I met a rider from Toronto who said he found a place to camp there that was only eight dollars for bikers."

The man harrumphs, tips a long drag. "That your bike?" He belches and waves the bottle toward my bike leaning against the privacy fence.

"Yep."

He stumbles around the table to get a closer look, catching himself on the fence. "That seat looks comfy."

Someone from the mobile home row distracts him before he can sit on it. He staggers away. I drag the table closer to his tent.

The beer drinker never asked where I am going or where I am from. In a way, it's refreshing that my journey didn't register as something special. Like me, he is in his own world. We just happen to share a rented table.

POSTCARD FROM THE ROAD 7/20/17

The Bridger and Gallatin Ranges reared misty from some 40 miles away, rimming the vast valley north to south.

The road spilled east, pulling me tighter to the mountains, but I could not see a way through the growing wall. The rocky shoulder of the Gallatins shrugged against the tree-studded slopes of the Bridgers.

The road took a sudden curve and exposed a narrow gap. "A gentle climb" assured my ACA map. And it was so, even with an elevation of 5760 feet.

Like a giant's three steps up to his entryway, Bozeman Pass was three tiers of elevation gain. At the top, the door opened wide to yet another vista...and a long, long descent into Livingston.

STATS

62 miles, max speed 40, ride time 5:01, avg speed 12.3, TOTAL 2277.06

43

Keeping up the Good Ride

Thursday, July 21, 2016
Livingston to Columbus, Montana

"Hey, if you can ride your bike across the country, you can race."

Intoxicated by the perfume of rubber tires, the girl shook her head. "Nah, I don't think so."

He flattered her. She knew the stories of Mike Walden, coach of world champion sprinters Sue Novara and Sheila Young, heart behind the Wolverine Sports Club. Truth was, she lusted after the sparkling ten-speed racers dangling overhead in the iconic Continental Bicycle Shop.

"Our club trains every Tuesday night at the Dorais Velodrome in Detroit. You should come out. I'll set you up on a bike."

"Really?" She lifted her eyes.

"Sure. We have rides almost every night. On weekends we race." He grabbed a measuring tape and edged around the counter. "How old are you, anyway?"

"Twenty."

He dropped the tape. "Twenty! You're too old. I thought you weren't even sixteen."

Funny, I completely forgot about Dain, the Warmshower host I met in Ovando. I could have stayed with him instead of sharing a picnic table with a drunk. Ah, well, I survived.

Back on the ACA route through a sleepy town before seven. With its historic downtown streets lined with two-story brick buildings that house breweries, bakeries, and art galleries, Livingston reminds me of Royal Oak, a trendy Detroit suburb. The aroma of warm bagels wafting in the street snags my nostrils. *Turn the bike around!*

The Montana Cup Coffeehouse & Bakery's wood floors creak a welcome. In contrast with the drunk last night, two women, one behind the counter and the other a customer, are eager to hear about my ride.

Oh my. Cream cheese oozes at first bite of the toasty everything bagel, perfect in its crusty, dense dough. I close my eyes. Food is heaven, I am sure of it.

The frontage road follows the south edge of the Yellowstone River, a dreamy decline in elevation. Big sky, big everything. The Crazy Mountains erupt from green fields to the north, towering over monoliths of round bales stacked three high and triple-sized square bales stacked six high. I want to stop and take a photograph. I want to keep riding.

I pinch myself. With a slight tailwind, I keep riding.

The Crazies. Some might say I would be a natural denizen here, wandering among these sentinels rising lone above the prairie. I miss my Midwest forests, but admit—the Crazies cast mist from my mind.

I am free and spin effortlessly. *This is why I ride.*

Somewhere east of Greycliff, a cluster of homes with just over 100 residents, a cyclist approaches from the east. These meetings continue to remind me of crossing paths with riders in 1976. No surprise, really, given the expansion of routes by the ACA.

Colin, from Canada, is on his first tour. Twenty-four hundred miles under his tires since he left home in June (not much more than me) with another 1200 to go—southern California by the first of August. "I made every mistake you can," he says. "I worried about making my goal until I made myself ride 120 miles a day for ten days. I shipped my camping gear home and I'm staying in hotels. Now I think I can do it."

"Hey, you're doing great." I think I'm old enough to be his grandmother. I encourage him. "I once rode from San Diego to Jacksonville, Florida in seventeen days. If I could do that, you can get to California in eleven days."

Two more cyclists appear in the distance. Colin clips into his pedal. "I've crossed paths with them. They're nice."

"Keep up the good ride." I get why he doesn't want to dawdle. Me? Today I don't mind. I wait.

Cody and Lydia roll up grinning and relaxed. On their second, long tour, they're heading to the west coast. "Tomorrow we are staying at a friend of my father's ranch," Lydia says. "We're taking a rest day and going horseback riding."

Meeting these young people, wanderlust fulfilled by bicycle, warms my heart. There is hope for the future.

And I am not alone.

Sixty miles into the blistering day, Reed Point wavers into view, one forsaken street of western-flick storefronts. Population fewer than 200. Okay, maybe I am mistaken about the size of towns on this route.

The Waterhole Saloon greets the weary traveler with a false-front billboard advertising cowboy cooking. I swing around and hitch my steel steed to the saloon's railing. Inside, the natural wood bar, tables, and logs walls radiate orange with recessed lights above the bar. Deserted.

Eyes adjust. Wait, there, a woman sits at a high table, her wrinkled face glowing from a computer screen. She's playing a game. She slides off her chair, shoves it against the table, and wipes her hands on a soiled waist apron.

"Can I help you?" she spits.

"Like to get something to eat. Your kitchen open?"

"Yep."

I climb onto a bar stool. She flicks a stained menu my way.

"What's an Indian taco?"

"It's a taco salad on Indian fry bread. It's huge."

"I'll have one."

The fry bread fills a dinner plate, peaked with a Crazy Mountain of lettuce, cherry tomatoes, black olives, ground beef, and shredded cheese. I feel like Bob Dylan, knockin' on heaven's door.

"You're doing pretty good." The bartender watches as she chips ice from a cooler behind the bar. Perhaps the zeal with which I attack the mountain softens her tone. "The other day a big fella came in and ordered two of those. When I brought one out he said, 'You really didn't make two did you?' Nope!"

I slide the cleaned plate away. "That was delicious."

"I'm proud of you." A smile.

Another advantage of riding a recumbent: no bent over position to constrict a full belly.

Cruising into Columbus, a bustling town of 2000 residents, a sidewalk sign in front of a shady park signals a Farmer's Market today. Reminiscent of Garrison Keillor's Norwegian bachelor tales, two men, wearing dirt smudged coveralls and white t-shirts, are bookends on a bench. Beards sprout from their jaw lines like roots. In unison they nod "good'ay."

"You here for the farmer's market?" I ask.

"Yup. Starts at 4:30. We're early."

I glance at my watch. It's 2:30, not bad for an eighty-mile pedal.

"Where ya headed?"

"Looking for Itch-Kep-Pe Park tonight. Riding home to Michigan."

"Well, that there's a ride. The park is south of town, 'bout half a mile. Come back for the market."

"Thanks, I will."

With a name that sounds like Kitch-Iti-Kipi, the "Big Spring," in Michigan's UP, how can I not stay here for the night? Itch-Kep-Pe is a free (donations accepted) county park nestled next to the meandering Yellowstone River. Several campers occupy sites, plenty more large and shady ones are available. I settle near the restrooms (no showers). I skip the Mac & Cheese Compleats and instead enjoy sautéed green beans and cauliflower, compliments of the Norwegian bachelor farmers.

After dinner a sandy path leads me through brush and over saucer-sized river stones. The lowering sun dances off the swift, wide river. I reflect on Andy's bad news: "My brother Steve might not even last a year. Makes you think about stuff."

Indeed. Like, should I be riding right now? Or should I be home with Andy? With Mom? Those pesky "shoulds" again. Death. Dying is something you do alone, even with someone holding your hand they can't go with you. I don't want to die without ever having done this ride. I think Andy understands. Does Mom?

Wading out hip deep, I slowly sink and lay my head into mother waters. Food might taste like heaven, but this feels like heaven. Aches and sorrow dissipate, coolness caresses my skin. A hawk soars. *Hey, Dad.*

If I lift my feet I may drift forever.

STATS
81.55 miles, max speed 31.5, ride time 5:29, avg speed 15, TOTAL 2358.64

Hitched to the rail at Reed Point's saloon.

44

Wind and Whistles

Friday, July 22, 2016
Columbus to Worden, Montana

The fragrance of machine oil, rumbling mills and lathes and grinders, the precision inferred by optical comparators. A family open house at her father's place of employment impressed the twelve-year-old girl. She became the only kid out of seven to follow in his footsteps. In 1977, in days of government mandates to hire women, the girl landed a job in a union machine shop, a competitor to her father's latest employer.

Her father's boss told him, "Well, we won't be hiring any pissers here."

She snapped, "I'll be the best-damn pisser he's ever seen!"

Her father smiled.

A hint of giggling and muffled voices. The unmistakable music of a freewheel rousts me awake. *My bike!* I tear out of my sleeping bag, ready to launch after the thief rolling my recumbent away.

Except. My recumbent leans against the tree trunk right where I left it. An overhead security light exposes three youngsters running, one pushing a clunker bicycle, their shadows stretching across the grass like Peter Pan's. My FitBit registers peak zone heart rate. Whew, gotta get it down before sleep can return.

In the morning, a man in a gray hoodie that matches his bushy mustache interrupts my fill-the-water-bottles chore at an outside spigot. I answer his questions without pausing.

"Rich, come here, I want you to meet someone." He gestures to a younger man exiting the restroom. "This gal is riding her bike back to Michigan." He leans in. "My name is Dick. Rich here is my son. We're taking a vacation together."

"That's nice." I set to closing lids.

"I'm from Arizona originally, now I live in Massachusetts. Rich lives in Florida."

I wedge my bottles into their cages and look up, edging to hit the road. Rich has the same steel-blue eyes as his father. The young man is quiet, all smiles but his eyes dart. *Is that Bunny sniffling in my ear?* I relax against the brick wall and rest my arm on my load. "You are a long way from home," I say.

"We like to spend time together exploring new places," Dick says. For the next twenty minutes, we share stories of the road.

Rich blurts, "I got hit with an IED in Iraq. I had a closed head injury."

"I'm so sorry. Thank you for your service."

Rich bows his head and shuffles his feet. "No problem."

"He's getting the help he needs." Dick straightens and claps his son on the back. "And he's going to be a certified diver."

Both beam. *Thanks Bunny.*

Goodbye mountains, hello plains. Pedaling along the Yellowstone River, a mournful train whistle echoes thoughts of Rich. His positive attitude and his father's pride inspire me. I view the widening landscape with fresh eyes, noting the splendor of trees scraping life from a rocky bluff, the nobility of farmers irrigating burnt prairie into green. What is a life well-lived? Converting challenges into opportunities? Or accepting them?

And what of my life? Born into white privilege, raised by parents who did their best, surrounded by love and a bit of grace, have I risen to challenges more than those I've brought on myself? I fought my own way in the world, standing on the shoulders of women who opened doors. Will my life add to the lift? I hope so.

Getting through Billings is a challenge. Busy traffic, busy buildings, busy bees. Saved by a sidewalk, I find air-conditioned respite (and opportunity) at the Gotham Park Laundry. More luck. A northwest wind kicks an assist for twenty-five miles into Worden, despite a harrowing cycle on a narrow, two-lane road with no shoulder.

The town of fewer than 500 people is wedged between the river and a rail yard. On a side street, I wave down a local man cruising on a four-wheeler. "I'm looking for the Lions Homesteader City Park."

"Go back out to the main road." He points to SR 312, the road I came in on. "Turn left and keep going. You'll see it on the right, over the railroad tracks. Can't miss it."

"Do you know Dan Krum? He's supposed to be the contact for staying at the park." I left several messages for him yesterday.

"Nope."

Lions Park is right where Mr. Four-Wheeler directed me. Gnarled old trees, green grass littered with thicker-than-thigh tree limbs, two ball fields, and what looks to be a concession stand with indoor restrooms. *Drat, they're locked.*

Another message left with Mr. Krum, guess I'm on my own.

Resting my bike against a picnic table, I consider tent space. The sky over the distant Bull Mountains turns ominous, not quite like the day on SR 200 when I got hit with rain and hail, but disconcerting nonetheless. A gust of wind chases sand across the diamond and slams my bike to the ground. And just like that, a gale force whips up and keeps a'whipping.

A shed offers a windbreak, but creaking branches overhead unnerve me. I shudder at the call someone might have to make to Andy—sir, your wife got taken out by a tree limb in our park. The dugouts are wind tunnels and even behind glasses dust scratches my eyes. Best option is a square of parched grass behind a three-foot plank backstop topped with chain-link fence.

The wind grabs and teases. Panniers and sleeping bag anchor tent corners while I battle stakes into the dry ground (no sprinklers here). "Toto, we are not in Kansas anymore." I raise my fist to the wind. "Leave my stuff alone!"

Hoping my tent stays secure, I walk across the tracks to explore. A gas station and convenience store a block east is closed. On Main Street, a bar and a grocery store huddle between abandoned storefronts. The store looks like it's been here since the west was settled, its warped, wooden screen door slams behind me.

According to my FitBit these last two days, I'm overeating for the first time this entire trip. I don't care, I'm starving. I grab Cheez-Its and a package of day-old cherry turnovers for breakfast. I've stopped cooking in the mornings to get a jump on the heat.

"Do you have a bathroom I could use?" I ask the cashier.

"At the back, turn right."

"Thanks. By chance, do you know a Dan Krum?"

"Nope, can't say as I do.

"I'm riding my bike through and he's supposed to be the contact for camping at the park. The restrooms there are locked."

"We're open until eight, so you can come here. But we don't open up 'til eight tomorrow."

Some small town this is, doesn't anyone know Krum?

Worrying about being blown to Oz, I spread my body heavy and wait for sleep, longing for Midwest forest windbreaks. A gust pummels. A stake holding one side of the vestibule yanks out and the door thrashes like a flag in a hurricane. I re-secure it. Two short blasts of a train whistle, followed by a long, and a fourth short blast, warn a train is crossing the road.

Wind and whistles delay sleep until well after midnight.

STATS
70.38 miles, max speed 38.5, ride time 5:12, avg speed 13.5, TOTAL 2429.08

Me, at my OD grinder, circa 1980?

45

Guidance Guises

Saturday, July 23, 2016
Worden to Hysham, Montana

Lost. And stuck, truck tires half buried in sand on the narrow country road somewhere in the middle of Clare County. Late November darkening, long before cell phones. The girl told no one about her impromptu trip north.

At least the gunfire stopped.

Headlights bounced, four hunters in a beat-up station wagon. "Need help?" Even with all of them pushing, her truck wasn't going anywhere. "Come with us. We'll get our four-by-four and a towrope. Our cabin isn't far."

The girl squeezed in between two in the back seat. Her mother would have had a fit. "You went with four strange men?" But it turned out fine. Wives bustled about in the cabin, visible through golden-lit windows.

It wasn't until later, miles away in a state forest campground, that she considered the risk she took. She didn't need her Chevy Luv pickup headlights to set up her tent, but she left them on. If she were still living at home, her father would have flipped the light switch by now. "It's ten o'clock, time for bed!"

"Looks like you've done this a few times," a man's voice startled her.

"Yep. Just a few."

His questions got her talking. Fascinated by her cross-country trip, "By bicycle!" he pointed to an Argosy travel trailer parked on the next site. "Come over and have a beer with me."

What the heck, the girl decided, things worked out earlier.

Seems the man needed to talk. The year was 1977. After Vietnam activist Jane Fonda attacked the Dow Chemical Company during a speech at Central Michigan University, Dow cut off its funding. He was a Dow executive. "This is where I come to decompress."

She listened long into the night. In the morning, the girl crawled out of her snow-covered tent, nursing a hint of hangover. The man saw her stir. "Hey, I'll buy you breakfast."

Bladder screams WAKE UP! Or maybe I only notice because the wind calmed, and the temperature dropped. I sneak out without my headlamp and find relief in a dark corner by the shed, taking my chances of being arrested for indecent exposure.

All is well. For a while. As dawn lightens, a different calling. The dark corner won't do for this job. Faithful TP roll in hand, I find a knot of thick brush along the railroad tracks and burrow in like a bunny in a briar patch. Only bunnies will see me here. Do bunnies poo in the brush? Why, yes, they do.

Time to beat the wind.

Mountains fade, heat rises. Ten miles in, a butte levitates stark above the prairie. According to a sign on the frontage road, Pompey's Pillar National Historic Landmark is the "last physical evidence of the Lewis and Clark expedition." I wonder if it's the last physical evidence of anyone. Straight ahead the pavement hazes to gravel. I turn right and cross under the highway into town, if you can call a handful of deserted buildings and a few shacks a town.

Dead end. Left is the only choice. A short climb reaches a fork in the road. Right is not a choice. The ACA map shows the route paralleling I-94 and the Yellowstone River. Left it is. Another dead end. Flummoxed, I turn around.

A gray flash hops out of nowhere and stops. *Bunny!* It faces the way I came into town.

"Did I take the wrong turn?"

Bunny sits, nose and ears twitching.

"Am I supposed to go back?"

No response. We both sit.

"Okay, I'll go back." Rolling away, I glance in my helmet mirror. Bunny is gone.

Well, look at that. The gravel I turned from was a mirage, the road *is* paved.

"Thanks Bunny!"

Sixteen miles later a crosswind whips up. Ah, well, at least the breeze helps to cool. Every now and again a gust grips the fairing and my speed drops four or five miles per hour. Other times the wind noise disappears, and I hear grasses rustle in waves.

A gentle curve aligns me with the wind. Speed lifts. I am power.

When I roar into Hysham (population 312), my planned stop for the night, I consider continuing to Forsyth. *What's another twenty miles with a strong tailwind?*

Then I meet Janet.

Filthy-kneed, wearing thread-bare gardening gloves and mud boots, the gray-haired woman is climbing onto a four-wheeler loaded with five-gallon plastic buckets. Behind her a stone-tiered flower garden bursts with red, yellow, purple, and white blooms.

"Beautiful garden," I say.

Her smile is as wide and bright as the river. "I was just watering and weeding."

"Is there a grocery store around here?"

"No, but we have the Friendly Corner at the edge of town." She points east. "Where ya headed?"

"Well, I planned to stop here tonight, but I have such a nice tailwind I'm thinking of going on to Forsyth."

"The fair's going on there, you could sneak in and camp with the fair people. No one would care. They have a city pool too, to get cool. Oh. Wait. It's closed on Saturdays."

A crowded and noisy county fair? Ah. No. "What about your city park? My map says it has camping."

"Oh yes. Turn right at Friendly's, the park is a few blocks down. It's nice, right next to our city pool."

Straight away to Hysham Lions Park, a flat, open field with an unlocked toilet building. Shade under a pavilion invites me to lunch (beef jerky, a tortilla wrapped hunk of melted sharp cheddar cheese, and a snappy Braeburn apple for dessert). Beyond the park, kids cannonball off the diving board at the city pool.

I give up trying to anchor my maps from the wind and call Andy.

"Finally, a tailwind. Made great time today. Not sure if I should keep going or stop here for the night."

No hesitation. "Stay put," he says. "Take a break. Last night I dreamt I had to come rescue you."

He is too funny. "Yea, it'll be quieter here and I could use a good sleep. My weather app says the wind will die down but not shift direction. I'll hammer tomorrow."

The Hysham pool is free. I opt for a swim and a shower.

"A rancher donated the pool years ago," Mendy, the facility manager, and today's lifeguard, explains. "He wanted it kept free for all the kids so they stayed out of the drainage ditches and rivers. They rebuilt the building two years ago."

After a few leisurely laps in the sparkling water, I doze on the hot cement deck. Wind, cool against my wet skin, carries me to Harlow, the street where I grew up on the west side of Detroit, where we spent summers swimming in the three-foot-deep, fifteen-foot diameter pool Dad put up in the back yard, where we warmed belly-side down on the cement driveway slathered in Crisco oil, browning like wild things.

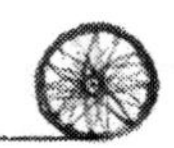

One day Dad and I were alone in the water. He wore a swim mask with his face in the water as he vacuumed the pool. Without warning, he lunged for the pool's edge, gulping for air in a dying elephant bellow. The same wake-us-up-wail where we'd find him on his hands and knees in the bathroom.

Mom said he had a hiatal hernia, that we weren't to worry. "The doctor said he'll pass out from lack of oxygen. Then his body will relax so he can catch his breath."

But I did worry, I knew what it felt like to fight for breath and hyperventilate until you passed out. Unbeknownst to us at the time, I had allergy-and exercise-induced asthma.

Too small to save him, I ran screaming to the front yard for help. It was the Twilight Zone with no one around. Terrified, I ran back, expecting to find him at the bottom of the pool. Instead, there he was against the picnic table catching his breath. His death throes alerted a neighbor who hurdled two sets of chain-link fences to drag him out.

Dad...

I peel myself from the pool deck and suffer a long, hot shower so delicious I don't notice it floods the room. Long narrow drains along the wall are blocked. Mendy catches me using a squeegee I found to direct water to another drain.

"That drain always gets blocked," she says.

"If you have a screwdriver, I'll clear it."

Mendy takes over the squeegee and talks. "I've lived here two-and-a-half years and I'm still an outsider. The pool board doesn't like me. The girls that work for me are daughters of board members. They say I bully them, but they just don't want to work, and they don't like me telling them to get to work. I took the weekend shift to find out what really goes on. Those kids told me the girls are always complaining about me."

"They had fun with you."

"Yea, they're fun. The thing is, there are state regulations. And the board doesn't want change. I work for the county commissioners, not them. They back me. The board can't fire me, but I'll resign if changes aren't made."

"You seem to do a great job. Can I write a letter to someone on your behalf?"

Mendy gets me paper and I pen a page-long missive. As I leave, she warns, "They come and spray for mosquitoes about nine o'clock sometimes. If you hear something that sounds like a helicopter, come hang at my house for a couple hours. It's the one on the corner with the deck."

I remember a fellow cyclist telling me how horrible it was getting stuck in their tent during a spraying. "Thanks, I will."

Relaxed, clean, and cooled off, I hang out in the wind under the pavilion and scheme my next couple of days. A couple long-mileage days and I should be out of Montana by Tuesday. Seems like it's taking forever.

The pavilion has electrical outlets, but the Lions Club is smart. Quarter coin boxes turn on the power. I'll need more than a couple dollars' worth to charge my phone, so I head to the Friendly Corner. The lady working the counter finds out where I'm staying. "Did you try the pool? I'm on the pool board."

I seize the opportunity. "It is the nicest city pool I've ever seen. And I've been in a lot of them. That gal running the place is great. She played games with the kids. After she closed up, she brought me the clothes I forgot in the shower. Hang on to that one."

I wish Mendy could see this woman's face. Wait till the commissioners get my letter.

And lucky me. No bug spraying.

POSTCARD FROM THE ROAD 7/23/16

Isaac told me the road between Custer and Hysham was rough. He struggled with its description. "It's broken up, uh...." he fumbled and finally just said, "it's not too bad."

I felt his pain. The wind suddenly kicked up west to northwest and he was heading into it. Isaac was riding from Philadelphia to San Francisco; we met in Custer at my first break of the morning.

"That's okay," he said. "I've been lucky so far. It's my turn today."

We shared tips on places to stay (or avoid) and route conditions. "Lots of cows," he added about the section of "rough" road I was about to meet. "A lot of cows. It was interesting."

When we parted and I turned onto the back road to Hysham, I thought I understood what he meant about the pavement. The macadam was rougher than the road I previously enjoyed, and he was right, it wasn't too bad.

For about three miles.

I crossed a cattle guard, which seemed a bit odd out in the middle of nowhere. They are common in front of ranch driveways and at the entrances and exits to the interstates. But out here?

The road beyond was "broken up" as Isaac said. Shake, rattle, and roll—no way to keep a good pace on this. Luckily, after a few miles I crossed another cattle guard and the road returned to rideable. The macadam almost felt smooth.

As for the cows? I first noticed cow pies strewn about the road as if a cattle drive recently came through. Nope. Turns out this is free range country.

"Nice cow! Good morning! Now you just stay right there while I

cycle on by. I promise I won't try and get a selfie with you," I chattered, weaving my way past more than one cow who decided that standing in the middle of the road chewing her cud was a lovely place to be.

STATS
59.43 miles, max 38.5, time 4:12, AVG 14, TOTAL 2488.63

Dad at the beach with me (left), sisters Sue, and Cathy (right), 1958.

46

Food Angels

Sunday, July 24, 2016
Hysham to Miles City, Montana

Before she died, the girl's grandmother recorded stories into a portable tape player. It was Rick's idea and a good one. The girl's favorite? Grandma overhearing the girl's toddler father cussing out a neighbor boy for jumping on the father's tricycle.

Grandmother's voice lilted, "And here was my little bratty-boy, he was letting off all these swear words... I gave him plenty of raps on the hind end...I told him he'd better not let me hear him say those kinds of words again. And I never did hear him. Never. Even when he grew up."

The girl thought this funny. "Grandma must never have been around Dad when he fixed something."

Or when he talked about segregation in the army: "Those fuckers, they made the black soldiers sleep in separate barracks. They bleed red just like we do, we're not any better than them."

Mountains no longer dominate the horizon. The Yellowstone River cuts a dark green swath through gold and sage fields rolling forth like a shaken magic carpet. South by southwest, the wind dries sweat into salt rivers across my face. I ride up every hill. Not too many weeks ago I'd be counting steps.

Mom, and her eighty-eighth birthday in four days, is on my mind. Last year I sat with Dad while she took the tour bus to the casino. Flowers? Banal. Chocolate? Might melt. Ah! Toenails. One of the things I worried about being gone all summer—Mom's toenails. I've been cutting them for years; they're gonna grow long before I get home. I take a break in Forsyth, last services between here and Miles City, tonight's destination some fifty miles down the road. A call the front desk at Park Place Senior Apartments gets me the number of the woman who gives pedicures.

"Just have your mom call me," the pedicurist says. "I can come anytime."

"How do I pay you?" Her service is $35. "I'm in the middle of Montana on my bicycle and won't be home for weeks."

The woman doesn't blink. Sure, daughters schedule pedicures for their mothers, but how many call from their bicycle in the middle of the country? "Pay me when you get home. That way, if something happens, you don't have to worry."

Her nonchalance at "if something happens" catches my breath. Dad didn't dub Park Place "God's waiting room" for nothing. "The only way I'm getting out of here is feet first," he'd say. I mail a birthday card with the pedicurist's contact information. Guess I'll have to wait to see what happens.

Practiced at stalling, I peek at emails and Facebook before getting back to pedaling. Bob Downes, the rider I met in Ovando, is blasting 100-mile days east on SR 200. He'll be in Glendive tonight. I'm still a day away, preferring to rest before dark.

My own pushed pace is not what I imagined. There was a time I would have jumped off-route without a second thought to explore, say, the "Enchanted Highway" in North Dakota, a thirty-two-mile stretch heading south off my west-bound route. True, I was on a deadline to Missoula when I passed it. I thought my return would be my usual what-is-down-that-road approach, I thought I might take the Enchanted Highway side trip coming east.

Now I know I won't. I am focused. Like the yellow-brick road to Oz, the wide green line representing I-94 on the Montana State map enchants me straight-away home. Imagine…this highway touches Michigan from here.

Andy.

Miles out from Miles City, a cut metal sign brands the town of 8,000 people with a rocking "MC." A herd of horses graze behind a wire fence. One stretches its neck toward a round bale left between the highway lanes and the fence. Only thing is, the horses aren't moving. Am I heat exhausted? Nah, more metal sculptures. *Random art next to the highway, wouldn't that be a fun picture book?*

I dig out my Nikon. Funny how little I've been using it. Maybe because this ride has become such an internal journey, I am more interested in absorbing everything instead of observing it through a viewfinder. Maybe writing has become more relevant than photos. It is what it is. I admit to having fun shooting this grouping. And when I turn back to my bike, kissed by yellow ditch daisies, I see the beauty in my steel steed.

Tour Easy. A most appropriate name.

In search of the Big Sky Camp and RV Park, downtown Miles City is an easy-to-slip-through ghost town. At a tee intersection, I turn right and get caught up in a suburban shopping zone. Traffic that wasn't downtown is here. When I see an interchange for I-94 with no sign of the RV Park, I realize I've taken a wrong turn. Turn around time.

I ride and ride. The road curves with a set of railroad tracks into the wastelands. Could I have missed the park? The map shows I'm heading the right direction. I keep pedaling, sweat drained. The road heads back toward the interstate. There, just before a rise to cross I-94, a plywood billboard heralds the Big Sky RV Park. The striated butte backdrop gives the camp a badland-feel, but a shed-sized wooden windmill in front of the office is disorienting.

Bonk-y I am, as well as scorched.

The office is cool; a pool table waits in darkness to the right of a cashier's counter and travel brochures line the wall opposite the door. Didn't the online description of the park include a convenience store? Here, only a hand-drawn poster board, duct-taped to a noisy refrigerator, advertises, "Pop. Water. Ice. Ice Cream."

A man enters from an inside door. I catch a quick glance of a couch and a Lazy-boy chair behind him. A television blares. *Must be the owner.*

"You can have any tent site along the outer edge." He points out the window to a grassy stretch of picnic tables under short shade trees. A pickup truck camper pulls in to block the view. A middle-aged couple stumbles in with a panting Brussels Griffon yapping around their feet.

"Do you have any other food for sale besides ice cream?" I ask the owner.

"Nope."

"Ugh, after almost eighty miles, I sure don't feel like riding back into town to find a store."

He shrugs.

Looks like tuna and instant mashed potatoes for me tonight. Ah, well. I register and step back to give the new campers a chance at the counter.

Before I can plant my rear end in the shade at the first tent-site, the couple with the yappy dog drives up.

"Hi, I'm Mark." The man thumbs to the woman in the passenger seat. "That's my wife, Renee. We're having hamburgers for dinner if you'd like to join us."

"That's the best offer I've had all day!"

They don't give me a chance to say anything more. As Mark pulls away, Renee leans over. "Come by at five o'clock."

"Come on in." Renee waves when I knock. She turns to tend sizzling hamburgers in a fry pan. "Sorry for the mess."

Clothes, bags of food, pots and pans cram every inch of space. Mark sits on a bench seat that butts up to an elevated bed over the cab. The Brussels Griffon hops with a vicious bark from his safe spot next to him.

"Willie, that's enough." Renee shakes her spatula toward the dog. "He doesn't like anyone but me."

I squeeze past to slip into the seat opposite Mark. He ignores Willie. "I'm retired from fixing gas appliances, but I still work," he says. "We bought this camper in Ohio to save on hotel costs for the drive back home to Washington. We've been visiting Renee's niece."

"Where in Ohio? My brother lives in Defiance."

Renee pauses with her spatula in mid-air and stares out the window over the stove. "Uh. I can't remember. I visit her every year. She's like a daughter to me. My sister, her mother, died of an overdose. It's a blessing it's over. I had to retire from nursing because of my arthritis. My eighty-year-old mother lives with me now and I take care of her."

"Renee was in a three-week coma over Christmas." Mark nods toward his wife, who assembles our plates with precision—hamburger on a bun, spoonful of baked beans, half-cob of corn. "She had pneumonia. And back surgery."

"I can't remember any of it." Renee pushes Willie off the bench and sits next to Mark. The dog jumps up and over her to the bed behind them. Most of her hamburger ends up in Willie.

"Care for an ice cream?" Renee extracts a softening ice cream sandwich from a small freezer.

"Sure, thanks." Having to throw my clothes into the dryer gives me an exit. I shoot a selfie with them before bidding a much-appreciated good evening.

I pass the park pool, a square puddle of chlorine-scented water surrounded by a six-foot chain-link fence. A young couple sits at its edge drinking beer. He is bare chested, with long, dish-blond hair; she's wearing cut-off jeans and a bikini top.

"Hey, where ya from?" the man asks through a few missing teeth. He lifts his beer in salute to my usual spiel, "We rode our bikes all across Montana. We pulled kiddy trailers with all of our stuff. One time we were riding along some railroad tracks and we had to get out of the way of a train."

The woman pipes in. "When we got here, I got a job at the hotel. We camped out all summer while I worked. In the winter, my boss gave us a room to stay in."

"We both work at the hotel now, and have a trailer here," he says.

A step up in accommodations. People survive.

My cell phone dies while I'm talking to Andy before bed. There are extra plugs in the laundry room. On the way, a yellow lab-mix barks and rushes out from behind a triangle-shaped pop-up camper. He gets yanked off his feet by a cable tied to a tree.

"Cody!" A woman exits the trailer. "So sorry, he's got manners to learn."

Cody wiggles like mad. I am not afraid. "Can I pet him?"

"Sure." He wiggles more madly.

"I'm heading over to charge my phone."

"You can plug it in here. Come, join us."

Verna travels with her daughter Hannah, who is barely fourteen. "We're from Wisconsin, heading home. We take a road trip every summer. I want to expose Hannah to things while she's still interested in being with me."

"Before you get annoying." I laugh. "Good for you."

"This is our longest trip so far. Six weeks. How long are you staying?"

"Just the night."

"We've been here a few days. Plan to leave the day after tomorrow." She ducks into the trailer to retrieve a foil container of hot lasagna. "Have dinner with us?"

"Thanks, but I ate with a couple that pulled in behind me today."

When Verna's mosquito coil can't keep the bombers at bay, we retreat into her darling abode. Cody curls up on the floor at my feet and we three gals giggle for hours.

"I hope we run into you again," Verna bids goodnight.

STATS
78.05 miles, max speed 35.5, ride time 6:06, avg speed 12.8, TOTAL 2567.59

47

Easier to ask Forgiveness

Monday, July 25, 2016
Miles City to Glendive, Montana

Four narrow lanes of bustling city traffic. The girl hugged the curb as close as she dared on her two-wheeled commute to work the afternoon shift. A quick glance in the homemade dentist-mirror-on-a-spoke attached to her helmet caused her to brace. A black 1978 Cadillac cruised into her space, its square fender catching the dropped handlebar of her Raleigh.

Time slowed. Scratching handlebar along the fender, the girl popped hip and elbow against metal to bounce from the encroaching car mirror and slammed right back against the passenger door to avoid crashing into a telephone pole. The driver's face froze in a shriek as she caromed the length of the car.

The girl exhaled. She didn't know how she avoided wiping out. "Guess I'll take my good luck with life and death on the road."

The sparkling waters of the Yellowstone River beg for submersion, but the road snakes too high on the hillside to reach cool bliss. Heat hisses like a rattler, poisoning my power. Nothing to do but drink, pedal, and keep drinking. With every uphill pedal, my stomach sloshes.

At last, a rest-stop oasis on I-94, twenty miles outside Glendive. Under the shade of a cement shelter, I ease back on my hot-air-filled sleeping pad, atop a cooler-than-air concrete picnic table. Cold water, dripping from my wind-charm bandana draped over my face, is a poor substitute for the river. I'll take it.

After a long nap, I meet Marvin. From Idaho, he heads to Fergus Falls, Minnesota for his sixty-eighth high school reunion.

"When I was in first grade," he says, "the teacher took two marbles away from another boy. She put them in a box by the pencil sharpener.

I wanted them marbles. I sharpened my pencil and snuck them out. At the end of class, the teacher went to give them back, but of course, they were gone. She demanded their return. But I was afraid, even afraid to give them back to the boy after school. Then he moved away. At our sixtieth reunion, I wanted to write a check to the guy, but he wasn't there. No one else remembered him. So now I'll give the school $500 to start a scholarship fund. The Stolen Marbles Scholarship."

Marvin takes a breath.

"I hitchhiked to Montana when I was fifteen to meet up with my sister's older boyfriend. Older, ha! Seventeen he was. My dad lent me $400 to buy his 1936 Harley. I was screwing around at a gas station and crashed into the pump. Remember those globes on top?"

I nod.

He continues. "It broke. I got the heck out of there, I tell ya. Seventy-one years later I went back. No gas station. The local chamber of commerce knew the owner's grandson. He opened another station right across the street. So, I went there. Found out the grandson's fourteen-year-old granddaughter wanted to sing. Well, I was a voice teacher. I gave the girl $500 for voice lessons."

His stories of restitution capture me. "You could write a book," I say.

"Well, I am writing a chapter to share with my grandkids. Things I've learned. You need to fess up if you've done something wrong."

Marvin carried these mistakes with him his entire life. His joy at closing the circle radiates, reviving me.

Glendive. From here I retrace my route west. There's the gas station where I stopped for directions to SR 200. The cashier couldn't help, two guys at the pumps filling up a work vehicle pointed me right. Here's the narrow bridge over the Yellowstone River. My passage held up traffic on the downhill curve, this time I take the sidewalk.

The ACA map lists a private campground and a state park beyond a steep uphill from the river basin, past the Subway where I enjoyed an early lunch, thanks to my friend Tammy. But. I am done. Luckily, the addendum sheet shows camping at Lloyd Square Park, right here in town.

The park sprawls around the city pool under soaring cottonwood trees. From the screams and splashes, I'd say half of Glendive's 5000 residents are cooling off. I pull up to a cedar sided Snak Shak. "I'll take a Gatorade. Do you know where I can set up my tent?"

The woman manning the Shak sets her phone on the counter. "I don't think there's camping here anymore."

"Says right here there's camping." I hold out the ACA addendum.

"Let me call someone."

I guzzle the Gatorade. Not good news.

"Yep, what I thought. The Rec Department says no camping."

I hang my head.

"But hey, I'm a rebel. I'd stay anyway. If anyone questions you, show 'em that paper. You might want to set up over there." She nods to a dusty hill behind the pool.

"Any public bathrooms?"

"That brick building." She points to the far side of the park. Flat, grassy, with a pavilion. A lot of steps in the middle of the night if I stay in the less conspicuous place she suggests.

Ask forgiveness instead of permission.

POSTCARD FROM THE ROAD 7/25/16

Sometimes the day's ride is just about survival. Like the fuzzy-eared fox at six-thirty this morning that darted across the freeway entrance as I rolled down the ramp. He startled as I shifted but hopped to safety in a dry gulch below grade. He paused to look back and spun in a circle as if he couldn't believe his eyes.

I don't need any more vitamin D. Big sky, no relief from the burning orb. I take an early start to gain some miles before the heat drains my resolve.

There's a mileage marker—drink. Okay, you've passed it—drink. I think I see another one coming up—drink. Almost to the top of this hill (love how western grades get easier right before the apex, unlike Midwestern hills, which seem to finish straight up)—drink. Ah, a downhill—drink. Oh man, now I gotta pee.

The good news about a headwind, it helps cool me when I douse myself with water (that now feels like hot tea).

Seventy-seven miles in by the end of the day. With any luck, this could be my last night in Montana. Bring it on, North Dakota!

STATS

76.75 miles, max speed 35.5, ride time 6:25, avg speed 11.9, TOTAL 2644.39 HOT (ninety-six degrees) and headwind.

Some Tomodachis, somewhere in Colorado, 1976. From left to right, Mike, Bob, Steve, Loree, Len, me, Glen.

48

Moving From Montana Soon...

Tuesday, July 26, 2016
Glendive, Montana to Medora, North Dakota

Snow in the streetlight glow turned the city-night into a snow-globe. A knobby tire track wavered through an empty parking lot. Flashing lights sliced the serenity red and blue.

The police officer lowered his window to a gust of fluff. "Are you running away from home?"

The girl, a daypack on her back, wiped snow from her bike's headlight and lifted goggles and balaclava. "No sir, I'm running home from work."

"Are you crazy? It's after midnight, aren't you afraid to get hit by drunk drivers?"

"New on this shift?"

"Why, yes I am."

"Every new cop pulls me over. Maybe you should stop those drunks instead."

Maybe I should have followed the advice of the Snak Shak lady and set up on yonder hill instead of here, where five-in-the-morning sprinklers add insult following a four-in-the-morning thunderstorm. Everything is soaked. A seam-sealed fly is no protection against high-pressure water jetting low. Between blasts, I rescue bike and gear to the overhang of the bathroom entrance, pull stakes, and whisk the tent to the pavilion which offers no protection from the wind. I retreat to the one splotch of dry pavement by the restroom.

Against the brick wall of the building, a cup of coffee warms my hands. In between spoonfuls of oatmeal I call Andy, confident with the time difference I won't wake him.

"Yep, I'm up." he says. "I see on the computer weather you've got some rain."

"The sky is crazy. Wind blasts from the east, but clouds race from the west. How does that happen?"

"Not sure, but the radar shows it moving through. You should be okay."

I smile at the image of him consulting his computer. "I hope you're right. Only sixty-five miles to Medora, so I can afford to sit a while. I'm leaving Montana!"

"You're doing good. How much longer to get home?"

"Three weeks? Gotta get through North Dakota." Andy's voice in my ear is like a camp without sprinklers, a world with tailwinds, and heat of a different kind. We connect beneath the raging storm. "I can't get Marvin and his stories of restitution out of my mind. Do you think it's a message for me?"

"What do you mean? You're not a thief, not that I know."

"No. I mean what I'm supposed to be getting out of this trip. Not restitution, maybe redemption? You know how I say no regrets? Make a decision and go for it, if it's wrong, learn from it and move on." I pause. Andy waits. "I've made my share of bad decisions."

"You're just experienced. And you know what I say about experience."

"Surviving failure. Thanks for that."

With a solid kerchunk, sprinklers disappear into the soggy grass. A second later, heads in the next zone burst forth like a litter of baby aliens. Spray bombards the shelter.

"Gotta go!"

No hope now of packing a dry tent, might as well head out. Nothing against this forever-wide state, but I'm ready to put it behind me.

Twenty-eight uphill miles to Wibaux and the east wind pushes the storm back to big-sky blues. Hello heat, I didn't think you'd leave me. In the Montana Welcome Center, a couple miles west of North Dakota, a matronly woman staffs a counter filled with maps and tourism brochures.

"Which way you headed?" She folds her hands over her denim dress.

"East. The wind is tough today, nice and cool in here. Is there road construction on 94?" I can't remember where I ran into that awful push-my-bike-up-dirt-hills section of highway. I made it through on a Sunday. Can't imagine fighting today's traffic.

"I don't think so, unless it's across the state line. Besides, I'd take

Old Highway 10 after Beach."

Ah yes, Beach. North Dakota, not Montana. "I took Old 10 on the way out, but I'm not looking forward to climbing that big hill at Sentinel Butte."

"Well, the wind is supposed to increase. The interstate is higher. You'll get protection going lower on the old road."

"That's a thought." And then there's Rick in his gas station community center. Be nice to stop in again.

If a west-bound rider passes me climbing the bench between Beach and Sentinel Butte, I will not look like the poor sap I screamed by twenty-seven days ago—the climb isn't as arduous as expected. I'm stronger now.

Rick waves me in. Big hugs. "How was the anniversary? Make it in time? Did you see anyone you knew from '76?"

Great. Yes. Yes. I serve myself an iced sweet tea and a pile of nacho chips and cheese from the back counter.

"What's the first thing you'll do when you get home?"

I swallow my immediate response (*grab my sweetie*) with a slop of cheese. Instead, this: "I'm looking forward to seeing if that mile-long, last hill home has flattened out."

Rick's blue eyes twinkle, as if he's read my mind.

Sixteen headwind miles into Medora is nothing.

The western-themed town nestles on the banks of the Little Missouri River at the entrance to the Theodore Roosevelt National Park. Its souvenir and western wear shops, ice cream parlors, and saloons sport wooden facades. Nothing touristy perks my fancy. My mind soars with eagles above the badlands. And comes to earth when setting up camp at the private Medora Campground.

My headlamp is missing from the small pocket in my tent.

Rote rituals in packing are crucial to keep track of things. Everything has its place. *Did it shake out when I bailed away from the sprinklers this morning?* Fluffing up my sleeping bag, I whisper a simple request (as Mom so often advises), "Saint Anthony, where is it?" And head for the showers.

Gonna be cool tonight. Ever try pulling on cycling tights in a steamy shower room? I shove my hands into the pockets to straighten them out.

What's this? My headlamp.

As I lay me down to read before sleep, "Thank you Saint Anthony."

POSTCARD FROM THE ROAD 7/26/16

The Montana highway department was nice enough to leave me a sign today. [Leaving Montana] Just so I would know…

STATS

64.12 miles, max speed 32, ride time 6:04, avg speed 10.5, TOTAL 2708.64

49

Specters

Wednesday, July 27, 2016
Medora to Glen Ullin, North Dakota

"Forty-thousand pigs?" Mikey's first words for miles.

The girl cast a worried look at her cycling companion. It was less than eight hours into an unsupported twenty-four-hour race through the sweltering, rolling farmlands of southern Ohio, and long past an unplanned break in front of a malodorous pig farm.

The farmer had greeted them.

"No, we don't need help, he just has a flat tire," the girl said. "You have a lot of pigs."

"Ya." He launched into a lengthy rendition of his operation. Unable to follow his thick, Amish-like accent, the girl turned attention to her friend. Head bowed, he sat in the ditch with his rear wheel resting on his knees, getting nowhere. "Here, let me fix it." The farmer kept talking.

"Forty-thousand pigs." Mikey shook his head. "No way that guy had 40,000 pigs! There were only, like, fifty little A-frame pig houses, not 40,000."

She laughed. "He said forty sows and pigs!"

Off early with a lazy-long climb out of Medora and back onto I-94. An east-northeast wind pecks my cheeks. Beyond high spitting clouds, morning glows golden beneath a flush of turquoise. Two suns appear. Through the haze, a mountain range hovers over slopes of flaxen crops.

I wipe my eyes and squint. There are no mountains, only open plains ahead. These North Dakota Badlands are up to their old tricks. There is only one sun: one sun creating crazy light against enchanting cloud formations, and bewitching me for miles.

Mostly uphill miles.

How can this be? I'm supposed to be losing altitude.

Ten miles out I leave the interstate to follow Old Highway 10. No shoulder on the fresh tarred and graveled road. A semi fills my mirror. Hello sage brush. Five miles per hour into the ditch is less risky than challenging a truck. After it roars by, I push my bike to the crest.

Did I stay on I-94 on this stretch? I still can't remember where that dreadful, dirt construction zone forced me to walk. Places mingle. Maybe I should have written more or taken more photos. What do I do with my time? All day, every day, I ride from one place to another. Oh. And find food, water, and a place to call home for the night.

Laurie Anderson still serenades. I might not be traveling at the speed of light, but as the day pulls me forward, energy grows.

Twenty-eight miles west of Glen Ullin, a cross soars skyward from the top of a knoll. Further on, a wind turbine turns in slow motion as if in reflection. In front of the contrasting pair, the modern campus of the Sacred Heart Monastery sprawls across the prairie, home of Benedictine sisters. The road curves. Traditional spires of the Assumption Abbey (Benedictine monks) loom over the town of Richardton, population 529.

Once a Catholic, always a Catholic. And every errant Catholic an ally. We've suffered the authority of nuns, the patriarchy (and sometimes abuse) of priests, sat in class from A to Z, lined up by height, bobby-pinned Kleenexes on our heads in place of forgotten beanies at mass, concocted venial sins in weekly confessions, kept two inches between the other sex, collected money for pagan babies, and wondered about a God who left unbaptized dead babies in an eternal state of limbo. Fear of hell hangs like a guillotine.

I often wondered if making us say the rosary together was Mom's way to keep us kids in check. Can't argue about who's not washing the dishes clean enough when we're praying. Can't hit our sister when we're on our knees rubbing beads. That, at least, I can understand. Even though I mentally checked out of the church at a young age, I obeyed my parents' Catholic house rules until I moved out. Wasn't worth the argument.

Mom surprised us all when she turned away from the church at age eighty (with Dad right behind her). "I'm dealing direct," she said. Her faith was true. She pained at the hypocrisies built up over the years, rules that kept her divorced brother from spending the night in our home, that allowed her birth-control-taking neighbors to place the Eucharist host on her tongue, that accused her and my father of being communists for raising independent thinkers, that exerted sexual control over her life when priests were molesting little boys.

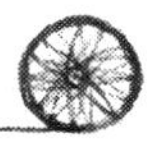

The specter of being raised Catholic is oppressive.

My legs pump like refurbished pistons but even if they had been drooping like last night's noodles, I will not stop at the state recreation area here. My mind is set, I steam past and keep on for Glen Ullin.

POSTCARD FROM THE ROAD 7/27/16

Red dirt collected on my tires as I rolled into the friendly Glen Ullin Memorial Park. The grandmotherly camp hostess had her one-and-a-half-year-old grandson in a stroller as she stood chatting with a gray-haired fellow tent camper. I overheard something about grilled hamburgers in town.

"The Lions Club has a cookout every Wednesday," she said.

After eight-seven miles I wasn't eager to ride another mile, but the idea of not having to cook and supporting a great organization at the same time was appealing.

"I'm going at five o'clock," the man she was talking to said. "You can ride with me in my truck if you want."

My tent went up in record time.

Jim is a beekeeper from Tennessee. He brought 200 hives to ND on the flat bed of his long truck a month or so ago.

"I'm retired so I can do this." Jim keeps a set of watercolors on the dash of his truck. He was an art director at an ad agency and later worked as a police sketch artist.

Interesting conversation and a ride into town. The least I could do was pay the six bucks for his burger.

STATS
87.09 miles, max speed 31, ride time 7:09, avg speed 12.1, TOTAL 2798.28

Mikey and me before a spring training ride, sometime in the mid-1980s.

50

To be Remembered

Thursday, July 28, 2016
Glen Ullin to Bismarck, North Dakota

Years later, the simple act of peeling potatoes still brought back an old friend's biting words, how he ranted to her about his newest employee, a girl barely out of high school. "What the hell is she thinking? She wastes the best part of the potato, using a knife instead of a peeler. What a dumb shit."

The girl didn't know why she was unable to challenge his attack. "Maybe no one taught her," she wished she had said. Perhaps she admired him too much, perhaps she didn't want him to know that she was a dumb shit too.

Every time she reached into her kitchen drawer, her hand paused over the peeler. Yes, a peeler makes sense. But she always drew a knife, each shave a small act of defiance, reparation for earlier inaction.

A flash. *Don't come crying to me!* A flare. *You had your fun, now suffer!* I swear, a pinch on the underside of my arm wakes me up. Mom? Nope, just a fierce thunderstorm. How long will my bladder hold? It's more than 200 steps to the outhouse.

Today is Mom's eighty-eighth birthday. I send an e-card. She must be up early too; she returns a thank-you card as quick as thunder. I FaceTime her from my sleeping bag.

"Happy birthday, Momma!"

"Too bad you aren't here; Cathy is coming to get me today. She's taking her computer in to get fixed and asked me to come along."

Oh good, she and Cathy are talking.

"After that we're going to Loui's for pizza."

Loui's Pizza. Yum. A bygone restaurant tucked between flat-roofs in Hazel Park, a mile north of the Detroit city limit. Flat-roofs are small machine shops. Now mostly deserted, they were booming businesses when the car companies were strong. Yet Loui's lives on. With its glittered ceiling tiles and Chianti bottles hanging from strings of lights around the walls, the place looks like its last remodel was in the 1950s. Waitresses who started working there in high school now push retirement age. Loui's is my favorite pizza place. Ever.

"That's great! Anyone else meeting you there?"

"I think everyone."

Except me. "We'll have another party when I get back, Mom."

"We'll see."

Thunder and lightning pass, but the steady rain feels like an all-day, good-for-the-earth rain. A call to Andy delays me getting wet a while longer. "Only two days into North Dakota and it feels like I'm getting home."

"I'm curious to see how you'll spend your time when you do."

"Well, I'm not going back to the newspaper job. I need to do my own writing. And I might need time before picking up my seventh puppy from Leader Dog."

"Good idea." Andy loves the puppies, but we both know the intense time commitment, especially the first weeks.

"I'm looking forward to our routine. You know, coffee in the morning at our nook by the back window, breakfast at the Sunrise Café on Wednesdays. Even Tuesday's bridge group at the senior center."

"Everyone asks about you," he says.

"When I was in Sentinel Butte, Rick asked me what I'll do first when I get home. I don't know why I didn't say my first thought."

"And what was that?"

"Grab my sweetie! I want to blast through North Dakota. I just want to ride. And get home. I might be alone on this trip, but I'm not alone in my life."

I think I hear him smile as we say goodbye.

What's that sniffing in my other ear? Bunny? *"Slow down, little one. You will get there soon enough and then you will wonder what you've missed."*

"Okay Bunny, I'll relax."

Twenty-six miles, and rain, spatter by until the ACA route exits the highway, six miles west of New Salem. More and more these Schwalbe Marathon tires impress me. Not one flat (I hope I don't jinx myself), even on the wet, debris-filled shoulder of I-94. Ahead, clouds break. At the top of a long, slow slog I pause. Over-my-head sunflowers leap from hills that roll up and away as far as I can see, bursting bright against the dark western sky.

I breathe in the view.

"How's this, Bunny? Might you suggest breaking out my Nikon to shoot this dramatic scene? Maybe not. Better experience in real-time, not viewfinder filtered?"

Another deep breath. Sunflower audience, facing east with me, presses me forward. In the distance a rounded butte rises from the prairie. From three miles away I read "New Salem" in white letters on its slope; on its top a forty-foot statue of a cow faces northward. "World's Largest Holstein Cow," the welcome sign declares. "Enjoy the view from Salem Sue."

I must not be the only one whose mind fatigues at this boundless horizon.

On a level stretch into Mandan, more than thirty-five life-sized metal silhouettes parade across a high ridge. A rancher on horseback leads a riderless horse and drives a herd of cattle. There are dogs, trees, a windmill, and a church. Like the metal signs announcing ranch names, the words "In Memory of Buddy Kahl," tower over the sculptures.

Why did I not see this when I rode west? Oh yes. I missed it because I took I-94 out of Mandan, instead of staying on Old 10. Like the lightning of this morning, the push-my-bike-up-dirt-hills memory flashes. The road construction was on SR 200 in Montana, not I-94 here. Jeez.

Who is this Buddy Kahl? I must remember to Google the name later.

"Buddy Kahl. K. A. H. L. Buddy Kahl. K. A. H. L. Buddy Kahl. K. A. H. L." Whispered down every aisle in Dan's Supermarket (where Howard left my lost rain jacket), spoken aloud with every pedal stroke into General Sibley Park.

"Buddy Kahl." I find his story on the internet. A friend to his community, the rancher died of a brain hemorrhage in 2003. His son constructed the 1200-foot-long sculpture. Loved and remembered.

Dad told everyone he met that his gift to the world was his children. "All seven of them are different," he'd say, "and they all contribute something to society."

Philosopher. Steward. She-Who-Paints-the-Sky. Peaceful Aura. Momma-Bear. Protector. DNA contributors all, except me.

What word for myself, beyond fringe ant, or student of life, or adventurist? Caretaker? Attendant? Dad's Pee-Meister, skilled with urinal positioning? Shower-Aide, Transporter, Elixir Mixologist? The word "equerry" rears up from an on-line Thesaurus. "An officer of the British royal household who attends or assists members of the royal family," the origin of which may relate to the Latin "equus" (horse). Ha, a horse I grew up to be?

Perhaps One-Who-Bears-Witness. How many hours did I sit with Dad, his eyes wandering the room, seeing people I could not?

In the end, Emissary. "Mom. He's gone." *But don't go to him on that side of the bed, you'll step in a pool of blackness.*

A lot to live up to. I hope we prove Dad right.

STATS
63.18 miles, max speed 39, ride time 5:11, avg speed 12.2, TOTAL 2861.54

Dad and Mom. April 10, 2015 on their sixty-seventh wedding anniversary.

51

Absolution

Friday, July 29, 2016
Bismarck to Gackle, North Dakota

The girl slammed the receiver into its cradle, stepped back from the pay phone booth hanging on the shop wall, and gasped for air. Stomping the length of the plant back to her grinding machine wasn't enough to disperse the anger. Anger with her mother. Anger with herself for letting her mother get to her.

She forced a block of carborundum against the side of a twenty-inch wheel whirling over 1500 rpm. Dust rose like smoke. The wheel wrenched the abrasive (and her thumb) into its fury.

No pain registered; capillaries cauterized, no blood flowed. The girl gaped at the open hatch to her pure, pearl-bone joint, it was as if someone scooped her flesh out with a spoon.

A hard lesson, letting triggers go.

In less than one hour of pedaling, a surprise.

"Bunny, look! It's the schoolhouse where I ate avocado-tomato burritos. I can't believe it was only twelve miles from General Sibley Park. It seemed like forever."

Weeks ago, in a rectangle of shade against the leeward face of the empty school, I didn't know how to continue, how to keep fighting that mother-of-all-winds. The west wind beat my brain. I imagined roadside wildflowers as aliens bending to suck out my life. That day's ride from Hazelton was a short forty-four miles, but I averaged less than nine miles per hour against winds twenty-four gusting to forty-five. At one point, it took two hours to crank out a measly 11.5. I thought: *I could walk faster*, and then a steep hill forced me to dismount and I learned I could not. Past that monstrous billboard—ABORTION KILLS A BEATING HEART.

"Oh Bunny, I feel so sorry for that thirty-two-days-ago me." The prospect of retracing those miles hangs like clammy green air before a tornado. "I've been skittish of this day, Bunny. Maybe it won't be so bad."

As usual, no response.

The road takes two ninety-degree turns up and around a butte, my thirty-two-days-ago Sisyphus push the other way. The savage billboard obliterates the horizon. A sudden punch in my chest does not slow me down, I veer and make the climb without breaking cadence.

Not for the first time, memory of my abortion grumbles.

Something wasn't right. I knew what it wasn't. It wasn't the minor foot injury that plagued me after a half-Ironman triathlon six weeks before a marathon, it wasn't the aftermath of an under-trained-for, five-hour marathon.

My period was late.

A friend (who hand built Lou's and my first tandem) stopped by for coffee one morning on his way to the bike store he owned. I worked the afternoon shift and was still in my jammies. I felt lousy. Sick-to-my-stomach lousy. My-life-is-about-to-be-fucked-up lousy. Excusing myself, I headed to the bathroom to check the pregnancy test paper.

Pregnant? Yes.

I remember nothing more about his visit. I told him nothing because he was not the one. (That one, a mutually convenient source of sexual release, would never know.) When he left I called Lou. She came to the clinic with me a few days later. Waking up in recovery I said to her, "I have never felt more relaxed in my entire life."

As far as anyone else knew, I had a D&C to solve a "woman" problem. Years later I told Andy, my then future husband. He hugged me tighter.

I made the best decision for me. No longer that misbehaving little girl who had to "wait 'til your father gets home," who cried "sorry, sorry, sorry" when Dad grabbed my hand to lead me to a spanking, I took it to heart when he said, "Don't be sorry. Just don't do it again."

I took responsibility for my carelessness and made sure it did not happen again.

In the mid-1980s, doctors did not want to do a tubal ligation on an unmarried, twenty-nine-year-old woman. Birth-control pills, yes, but a permanent solution? No. Finally, an OB/GYN took me at my word. "If you say you never want children, I believe you." When you find a doctor who respects that only you can make decisions about your body, never let them go.

For a while I counted the years. "My kid would have turned five this year," or "Let's see, nineteen this year." I suppose I wondered if or when remorse would ever hit.

Eventually I quit counting.

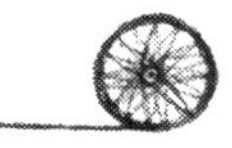

Rounding the second turn, the road drops like an elevator with a broken cable, a runaway dash stretching further east than I can see. It is the fury in the wind, no, that causes my eyes to water so?

"I am so sorry little baby." Wailing as much for her as for the thirty-one-years-ago me, I pull off my sunglasses to wipe my eyes. My gut expected some kind of backlash through this turn, but breaking down after all these years? I pump a fist into the sky. "Is *this* what it's all about?"

Just then, a hawk alights from a telephone cross-pole and swoops low in front of me as if to say, "Hey, I am here!" My gaze soars upon its wings. I sweep by another pole. A second hawk arises. A third pole, a third hawk. And then four! What are the chances?

"I am here!"

My body convulses strong enough to cause my stable recumbent to shudder.

"Dad! I am *so* sorry!"

The horrific image I carry since the Friday night he died last August swamps my brain. At 3:00 a.m., the sound of dripping dragged me from sleep. I ran into his bedroom, thinking he pulled out his catheter. Nope. I looked up. Blackness poured out of his mouth like sludge, down his white t-shirt and sheet, onto the white carpet. I gathered the sheet to dam the flow, leaned forward, and patted his arm.

"Dad."

He exhaled, like a sigh.

Once, and no more.

The blackness kept streaming.

I can't erase it. This, and the terminal agitation he suffered the long days prior. "Help me!" His arms jerked as if he were falling from a great height. Hours and hours, I mixed crushed Ativan with morphine, injected the cocktail against his inner cheek, and called the hospice nurse for instructions when no relief came.

The nurse said, "Sometimes when they finally settle, they pass."

There was nothing else I could do.

He settled after midnight, his breathing relaxed. Mom snored. I left his side to get a bit of sleep myself. The nurse told us she didn't think he would last through the weekend. I couldn't last myself without rest.

"Dad. I am sorry." I sob, remembering how Mom peeled his fingers from the bed rail to remove his wedding ring.

A hawk shrieks. It circles, calling again and again, soaring in a moving draft that seems to be following me like the cloud above Joe Bfstplk's head in the L'il Abner cartoon.

"What do you want? Is this what it's about?"

Shudders calm, the gruesome image gone. Instead, my mind's eye sees Dad reach toward me with his fingers and thumb to pinch my nose, just as he did the afternoon before he died. For a brilliant moment I am that little girl again, who got her nose stolen, not spanked.

I feel him with me, around me, in me.

"It's okay Pat." His voice echoes in memory. "Thank you, Dad," I whisper.

I coast a good long time.

The sight of a Medusa-woman in the driver's seat of a compact SUV parked on a dirt path, her hair wild, golden fuzz, waving like mad as I zoom by, a magnificent Golden Retriever smiling out the open window, is a dream. A few miles later, a cyclist stretching over his bike in the middle of the opposite lane next to a white minivan on the shoulder, slams me into the now, the now that looks back with compassion and understanding on the thirty-two-days-ago me. And the thirty-one-years-ago me.

The rider waves me down. "You passed my wife back there!"

"The one with the Golden?"

"Yes, she texted me that a recumbent was coming this way."

The man is riding across the country, east to west. "My wife wouldn't let me go alone, so she's sagging." Another coincidence, his wife is raising the Golden to be a service dog.

The driver of the mini-van is part of a four-person group riding east. "We take turns driving the sag vehicle." Now I notice the bike on a car rack.

With miles to go, I beg off. "Safe travels!"

I reach Hazelton before lunch. At a corner gas station and convenience store I meet Loren and Robin, two twenty-something young men riding from Washington state to their hometown of Minneapolis.

"I went to school for a year out there," Robin says, "but I dropped out to ride home. Loren here took the train to ride back with me. I've toured some, but this is Loren's first."

"It's great!" Loren says. "Once you tour you get the bug and just want to keep doing it. I got hit by a car in Bismarck, though. I'm okay, but my wheel got bent. Took it to a local shop, I couldn't believe the mechanic banged my rim against a workbench to straighten it!"

I nod. Done my share of banging bent rims. I don't have the heart to tell him of other brutal repairs during my time with the Rover, like re-tapping and Heli-coiling stripped crank arms to reinstall pedals or filing a bottom bracket shell square for a new bearing. The bottom bracket is the frame part where the crank attaches.

"Where ya headed today?" I ask.

"We're just getting going. Stayed here last night, heading for the Honey Hub."

Gackle, close to seventy miles away. A late start. Hung over is my guess. Good for them, it's fun to be young and on an adventure.

"Hey, if you help me make it to Gackle, I'll look at your bike when we get there. You don't know this, but I'm the best mechanic you know."

Traffic out of Hazelton is nonexistent so we ride abreast. Chatting with someone other than Bunny is a pleasant diversion. A rider pulling a trailer sneaks up behind us. Bruce, from Ohio, slows to gossip. I swear I see a puff of Road Runner smoke from his wheel as he resumes his pace to disappear in the distance.

Ahead, the prairie rolls on.

My long wheelbase, fairing, and heavier load gives me an advantage over Loren and Robin on descending grades along North Dakota State Highway 34. We don't stay riding side-by-side for long.

Two and a half miles from Napoleon, the road curves between two of three lakes, all named McKenna Lake. Several abandoned cabins stand windowsill deep in the water yards from shore. Further out, a line of bleached, barkless trees rise like sentries. There were white caps when I passed by riding west. Even with storm clouds on the horizon today, the wind is not as strong. To kill time, I stop for a photo.

My strategy works, the guys catch me up. "I broke a spoke," Robin says.

"Let's take a break in Napoleon and I'll fix it. Do you have a spare?"

"Nope."

"No worries, I have a FiberFix spoke."

Less than a mile off-route, Ken's Shopping Center provides a convenient stop. We buy snacks, refill water bottles, and empty ourselves. An overhang offers protection from a quick-to-rise thunderstorm while I fix Robin's spoke.

Neither of my companions is familiar with this emergency gizmo. While always a good idea to carry spares, spokes most often break on the drive-side of the rear wheel. Tools to remove the gears for access are too heavy to carry. The Kevlar FiberFix threads into the hub and rim without tools. One time I replaced four broken spokes with these kits for a rider on a five-day tour in Michigan; he rode three days with them. By the time I true Robin's wheel good enough for him to ride again, the rain stops.

Gackle. Ecstatic and vindicated, I clock over 100 miles—what took me two days to ride west a month ago. Don't need no stinkin' tailwinds!

Thanks again, Dad.

Today the Honey Hub looks like a circus. Tents dot the lawn, clothes hang on the line, riders mill about. Besides Robin, Loren, and me, Bruce is here, a young couple heading west, and a man with two teenage boys riding from the west coast to Washington D.C.

"Who wants pizza and beer? Put your share in the donation box." Bruce heads to the bar with his stripped-down trailer.

The westward couple picks up fallen apricots in the yard like they are

Easter eggs. "He's a forager," the woman says of her companion. The two are quiet, preferring their own company.

The man, Steve, owns a bicycle tour business based in Pennsylvania. "We offer guided tours, for teens mostly, but also adults and families. I had two more riders and a leader for this tour, but it all fell apart. Matt and Everett were the only two left, so I took them myself." The boys, both sixteen, hail from New York.

I tell Everett, the smiling one, "I'd have killed to do what you're doing when I was sixteen."

"I think it might be better at twenty," he replies, after my story about B'76 and the reunion. "It isn't what I expected. I thought I would ride all day until I was exhausted, and then I'd be somewhere. I didn't realize the logistics all the time, how you might have to stop before you are ready because there is nowhere else to stay. And shopping for food!"

I'm glad for the distraction of comrades but linger at the edges of conversations, Bunny-like, still and calm, and watchful.

STATS
109.32 miles, max speed 39, ride time 8:33, avg speed 12.8, TOTAL 2970.89

Leaving Hazelton. Another hawk! Photo by Robin Heil.

52

Allies

Saturday, July 30, 2016
Gackle to Enderlin, North Dakota

Mother told her
"Little girls can't grow up
to be horses."

why not?

long before she understood
double entendre—put something great between your legs—
she learned
mysterious self-pleasure
with a shimmy up the swing set pole

long before she understood
innuendo—you're growing up to be quite a nice young lady—
she learned
shameful forcing
from an uncle's foul tongue in her mouth

she felt the power in her legs
chasing imagination on a steel steed, ally
against the pressure of predators

Morning breaks cool with a south breeze. We head straight east on SR 46, supposedly the longest straight road in American. Yes, we. Robin and Loren, Steve and the boys, hoping to tackle the more than seventy miles of limited services all the way to Enderlin. Steve is sure we'll find camping, even though the ACA map lists nothing.

We leave together, but like yesterday, my downhill speed steals me ahead. "If you see my bike lying on the side of the road, don't look for me. I'll be taking a pee."

Pedaling shakes the tightness out of my legs like beating dirt out of a rug. I pause a few times to snack. No riders. Each time I press on, the wind kicks up a notch, I hesitate to hesitate when it shifts from the east. *Where oh where are those west winds?*

Short relief on the descent to Little Yellowstone Park. The park has more visitors than when I stayed here last month. My appetite roars. I eat yet again and charge my phone in the pavilion. No riders, the wind must give trouble. *I guess I'm on my own.*

After a climb up from the watershed, corn and wheat fields stretch into the horizon like a Great Lake. At least a half-mile away, a white blob flutters on the shoulder. The mysterious object? A plastic grocery bag. Three small bags of Cheez-its and two white bread sandwiches kept it anchored. Dropped by a cyclist? I snap it under bungee cords and keep riding.

About ten miles west of Enderlin a soft thunk and the plastic bag slips free. Rolling thunder stops me in my tracks when I turn around to retrieve it. *What the heck?* A sorrel mare, speckled lightning blazing her face, gallops toward me full bore. In a split instant, I notice mere strands of barbed wire between us. The freight train stops with finesse. She bobs her head to stare at me.

"Good afternoon!"

She bolts. Races along the fence line, turns and charges back, nods, then gallops across the overgrown field. She runs hard, circling rusty red farm equipment and a wood-railed trailer with broken boards. She never takes her eyes off me.

I can't take my eyes off her. Despite visible ribs, the mare is muscular grace.

Straight at me again, again she pulls short at the fence. She snorts and shakes, her mane quivers. Speeding off she shows no sign of quitting. I snap pictures. She doesn't disappoint, she leaps with all four feet in the air, bucks, and sprints back to face me.

Her head stretches over the wire, luring me to the fence. I extend my hand. She sniffs me and exhales, her breath feels like life itself. With a wink, she turns her head for a portrait.

"Are you trying to tell me something?" She doesn't bolt this time. "Or are these flies getting to you?" She suffers me to brush her cheeks and neck.

I linger a long while. She seems to enjoy the quiet between us as much as I do. Eventually, she trots to the far side of the field. She watches me repack my camera, watches me lift my bike to mount.

I am not a horsewoman, but this creature speaks to me. *"Take me home."* I am loath to leave her.

As if this horse has kicked me into another dimension, my cyclo-computer reads crazy, first thirty-two miles per hour, then three, then no speed at all. It started acting up yesterday after leaving Hazelton. A wiggle of the mount and it worked fine the rest of the day. I'm not stopping now, I'll look at it tonight.

Enderlin is deserted, except for three youngsters playing on the sidewalk a block from the main drag. The kids wave. I'm guessing they are too young to know if, or where the town allows camping. A skinny woman steps out of a bowling alley for a cigarette.

"Do you know where I could camp here?"

She coughs. "There's the park by the highway, but I'd stay at the Lions Park here in town."

I recall the park coming into town. Picnic tables and a rickety shelter, no restrooms. She's right, the Lions Park by the library (where I enjoyed a shady lunch on my way west) is a better choice. Lush level grass, plenty of shade, flush toilets, and a pavilion with electrical outlets. Oh, I hadn't noticed the busy rail yard across the road. Ah, well. Nothing is perfect.

I check my computer. The magnetic pickup on the front spoke is aligned and the harness is intact. I remove the head and snap it back on. The screen shows information, so the battery is fine, why won't it register speed? I don't want to reboot, the memory will clear. Pushing buttons I find the screen with the wheel size: 157. Good thing I've been recording daily stats.

Deep breath. Pop the battery out, put it back in upside down, remove and reinstall. The wheel size is still correct. Good. Test. Still no speed. Argh. Not the end of the world, but I will miss watching my speed and total mileage.

I set water to boil and check maps with the help of my headlamp. If I divert south of Fargo to head east to Pelican Rapids, I'll be out of North Dakota tomorrow, a day faster than when I rode west. And a day closer to Karen in Eau Claire. She follows my Facebook posts. "If you can get here the weekend of August 6-7," she emailed, "you can come to puppy camp. Everyone would love to see you."

Puppy camp is a Leader Dogs for the Blind training weekend Karen holds for Midwest puppy raisers. Deb Donnelly will be there, puppy counselors I know, and at least twenty raisers and their pups.

An intermediate goal.

"The cavalry is here!" Steve soars in with plastic grocery bags hanging from his handlebars. Matt and Everett land in formation behind him. Smiles all.

"I figured you guys stopped at the Little Yellowstone Park. That wind was tough. Loren and Robin texted me. They headed north to Jamestown to get out of it."

"We fought the wind two miles into Marion for bar pizza. Boy, was that fun coming north with the wind behind us. Awesome sauce!"

Amazed they made it, I am surprised how happy I am that they did. Rabbits, hawks, horses, and strangers keep away loneliness; still, companions passionate about two-wheeled adventures are delightful.

We eat together in the dark. The trains stop rolling in. Crickets sing.

STATS
So – last total 64.26 (+10 = 74.26) miles, max speed 33.5, ride time 5:09, avg speed 12.4, TOTAL 3025.18 + 10 = 3035,18

Spirit horse.

53

Like Being Reborn

Sunday, July 31, 2016
Enderlin, North Dakota to Pelican Rapids, Minnesota

They did it. In twenty-four hours, her father rode 115 miles, she rode 379. Together they claimed the father/daughter award. Winning overall woman was like eating moist carrot cake after licking clean the cream cheese frosting.

The next morning a phone call disrupted her reverie.

"Mary is dead." Hard to make out the words through her uncle's sobbing. "They found her in the bathroom at work. I couldn't get through to your dad, can you call him?"

She thought pedaling for twenty-four hours was hard. It was nothing compared to telling her father his only sister dropped dead of a heart attack at age fifty-six.

What wonders a lovely sleep (and no sprinklers) allow. Fresh eyes on my cyclo-computer find the harness wire stretched tight. Adding slack, I notice the connections on the mount are out of sync. A quick snap into place and abracadabra, it works! Why, oh why couldn't I have figured that out before clearing the data? Ah, well.

Steve sees me boiling water for oatmeal. "Why don't you eat with us at the convenience store on the way out of town?"

"Seriously? At the gas station on the corner?"

"Yes, they have great breakfast sandwiches you heat in a microwave. You never had one? Gotta try the egg and sausage muffin. Awesome sauce. My treat."

We breakfast on round cement tables outside the station. Glad I ate the oatmeal. Steve spreads out maps. "What's your route plan today? We'd like to get out of North Dakota."

"Me too. Shooting for Pelican Rapids. I'm not going through Fargo.

See how SR 46 goes east over I-29? Looks like there's a road that turns to Wolverton and then straight east. The ACA route through Minnesota is great—120 miles of bike path." I fill him in on my deviation through Wisconsin and the ferry across Lake Michigan.

"Sounds good." Steve nods and folds his maps. Without it being said, we become traveling companions.

The wind kicks up. I shake my head, where are the tailwinds? Cresting out of the Maple River watershed, a smooth downhill pulls me ahead. Thank you, Phineas! The construction driver promised a nice road the next time I came through.

Riding west in early summer I had no idea what plants were sprouting across these sprawling fields. I miss Lou behind me, my tandem horticulturist. "Those are sugar beets," she might have said.

Nope. Not sugar beets. Sunflowers! Will shooting photos give the guys time to catch up? The ocean of yellow faces curve against a deep green edge of trees in the distance. As I tramp across a berm of waist-high grass to enter the flowery surf, the horizon disappears. Overhead, only blue. Leaves rustle like fans blooming joy above the cracked soil. Can a sunflower smile?

I could get lost out here.

No sign of the others. Can't keep waiting, we must ride our own ride.

Kindred, kind Kindred, calls to me where the Northern Tier turns north toward Fargo (where I had a police-protected lunch), but I know the dangers of stopping. This wind. I continue, only to pause where SR 46 lifts over a highway. A bing of my phone interrupts a juicy bite of Braeburn.

Steve texts, "We will stop in Kindred for lunch. Struggling with the wind."

"Yes, not fun. I'm just crossing 29. Hoping it's less than ten to Wolverton. I'll let you know. Mileage here is thirty-eight. Good news, no hills. Cross 29, turn right onto 81, after four miles turn left at Christine onto 2N."

"Sounds good. Wind any better?"

"Sorry. No."

"It'll get better." I suppose helping two young riders through a challenging day demands a positive attitude.

Onward. Wide-open flatness in every direction, wind on my face is a blast furnace. A spurt of trees like broccoli stalks announce a distant town, much like Kansas grain elevators did in 1976. I remember pedaling all day long and the things got no bigger, we thought we'd never get there.

What's left of my water is warm. Christine is a few houses with shade trees, nothing more. And no cover for bladder relief. Fields stretch into haze. The road dips to cross the North Bois de Sioux River and rises to deliver me into Minnesota. *Goodbye North Dakota.*

An update text to Steve. "You'll enter Minnesota when you cross the river. Turn left onto Highway 30 in Wolverton. From there it's about thirty-two miles to Pelican Rapids."

"Awesome sauce. And thank you!"

Wolverton is bigger than Christine, there must be a gas station, at least. No. Another abandoned town next to a railroad track. Did I spy a water spigot on the side of that fire station? I lean my bike in a shady spot, grab a bottle, and walk back in search of water.

No spigot.

Across an alley, two women and a man sit at a round table under the shade of a large umbrella; a black, gray-faced dog lumbers next to a man watering flowers with a hose; two young kids play in the dirt. The sitters stop talking when I walk up. They embrace dripping glasses filled with ice and a bronze liquid.

"Excuse me," I croak through a parched mouth. "I'm riding through and I wonder where I could find a bathroom and a place to fill my water bottle. A store or gas station?"

The sitters laugh. "You won't find anything open today."

The dog wanders up and I squat to give him a pet. He leans into me.

"Can I fill my bottle with your hose, then? Sure is hot today."

The woman hesitates. The other two glance at her. "I guess."

The man stops watering, "Let her in to use the bathroom. And fill her bottle inside."

The woman glares back. If she held the hose, I wonder if she'd squirt him.

"It'll be okay, just show her in." He nods to the door.

The woman smacks her drink onto the table, heaves herself out of her chair, and motions me to follow her up a few wooden stairs. A little girl skips in behind us.

"Fill her bottle," the woman directs.

I hand it to the girl. "Thank you so much."

"The john's down here. Sorry, we don't have a door. We hang this up." She drapes a blue fleece blanket across nails in the wall above the opening and turns away.

I make it quick. The girl hands me my bottle, dripping with condensation. "Now I won't have to desecrate the corn fields." I lift the heaven-sent water to my lips and guzzle.

Back at my bike, I plant my butt on the sidewalk with my food bag and lean against the warm brick of Wolverton's locked Post Office. Halfway through my melted cheese wrap, a girl about fourteen strolls by with two ankle-biter white dogs on retractable leashes. They snort and wiggle right onto my lap, sniffing face and food.

"Hello puppies." I lift my sandwich out of reach. The girl stops at my feet. "Hi."

"Hi." She looks down at her flip-flop feet. She wears tight short shorts and a tank top. The wind blows her curly brown hair into a tangle around her face. She juts a hip, holds both leashes with one hand, and reaches up to brush hair away from her face. Her expression is pained, as if she stormed away from a family gathering using the dogs as an excuse to leave.

"You okay?"

"YES." She looks across to the railroad tracks and offers nothing else. I say nothing in return.

Time shifts. The wind keeps blowing. We are forever caught in this scene, only the dogs are energized.

She breaks the spell. "I guess I should get going." Her flip-flops echo against the pavement. The dogs linger, leashes trail until one more step yanks their collars. They tumble off my lap, gather feet, and pounce after her.

I should get going too. As I re-tackle the wind, a teenie bird flits ahead, chirping a cheery song. I sing along, "Keeeeeep, keeeeeep, keeeeeep! Keep pedaling' on!"

Traffic surges on the narrow, shoulderless road nearing Pelican Rapids. My left eye keeps glued to my helmet mirror. My phone rings. And rings. By the time I get off the road it stops. Mom. I FaceTime her back.

"Patti, where are you? It's late and I haven't heard from you yet."

I glance at my watch. It's after 7:00 p.m. "I'm in Minnesota!"

She sags back into her Lazy Boy chair. I love that I can see her where Dad used to sit, their jeweler's clock ticking on the wall above her head.

"I'm fine. Just a long, windy day. Again. I'm not too far from my stop."

"Where will you be tonight?"

"Pelican Rapids. I stayed there heading west. Remember when we FaceTimed and that guy on a trike said hello?"

She doesn't remember. I tell her about Steve and Matt and Everett. "I hope they make it here too."

"Well, be safe. I miss you. I love you." She blows me a kiss. When I blow a kiss back, "Love you too," her face brightens.

Before I return to play with traffic, I text Steve a warning. "Tough ride. After you cross 94, only twelve miles but the hills start. And traffic. Be careful!"

He texts back, "We only made it to Christine. We may stay here tonight. Grabbing some dinner right now."

"Was there a place to eat there? Probably a good plan. I am still eight miles out."

"Bar. Frozen pizza."

"Aw. Sorry the day is so tough."

"No worries. The boys are disappointed. We all wanted to play Bananagrams with you. Maybe we can catch up tomorrow. Where are you staying tomorrow night?"

"I'm shooting for Brandon. Chippewa County Park. A couple miles off route, but a nice park on a lake with showers. Depending on the winds tomorrow you might take 81 down to Wahpeton and then left to Fergus Falls. The bike path starts there. Good luck! Tell the guys I'm disappointed too."

I make it to Sherrin Memorial Park well before dark. The spot next to the pavilion where I camped on my westward jaunt is available. As I'm staking down my tent, a tattooed man holding a beer and dressed in nothing but wet swim trunks approaches. A tanned boy with jet-black hair, also wet, hovers a few feet behind him.

"Hi! Where you going?" Spanish accent.

"I'm riding home to Michigan." The man's eyes widen. "I rode to Missoula, Montana and now I'm heading back."

"Wow! That's fantastic! Miguel, did you hear that?" He gestures the boy to come closer.

Miguel Jr. is Miguel's son. His family commandeers a picnic table near the pedestrian bridge that connects Sherrin Memorial with E.L. Peterson Park over the Pelican River. Kids jumped off the bridge when I was here last; they must cool off this way too.

"How's the water? I see the pool is closed on Sundays. And the park shower is icy."

"It's great. You should jump in with us." Miguel Sr. drags me to meet his wife and daughters. His youngest daughter, Karina, is also a jumper.

"Oh, I don't know." I might want to try swimming, but jumping? Not so much.

"Oh yes, you have to try it. We'll jump with you!"

Miguel's wife encourages me. "My daughter will video tape you."

Before I know it, I take off shoes, socks, and Fitbit, empty my pockets, and walk onto the bridge. Following Miguel's lead, I climb onto the narrow metal railing.

"Grab onto the wire."

I look down. It feels like standing at the top of the Mackinac Bridge, 200 feet above the straits. My knuckles whiten. Only about twelve feet, twelve feet too far. My feet will slip on the railing, I'll do a Greg Louganis, banging my head against the rail on the way down to knock myself out. My heart closes off my windpipe.

"Are you sure it's deep enough?" I remember the feet of a childhood friend standing straight up after diving headfirst into our backyard swimming pool. The girl was okay, but the memory of a classmate breaking his neck after diving into his family's pool from their garage roof made the sight horrific.

Miguel has none of it. "It's just water, it'll support you." His smiling face is the star around which I revolve. *How is it I trust this complete stranger with my life?* "Okay," he says, "on three we'll all jump."

Miguel Jr. nods, ready to fly. Karina touches my arm. "We're with you."

It takes eternity to hit the water. No. I hit the water before I even jump off. Water slows the death grip of gravity, I descend in darkness, a cool paradise of equilibrium.

Am I being reborn?

Bubbles raise me to the surface. My lungs inhale a reverse scream. Miguel and his kids slap water, reach for high-fives. I pump my fist.

"You did it!" Miguel yells. The shore crew cheers like we scored perfect tens.

Water caresses and holds my body. The wind and sun and heat dissolve. Miguel Jr. directs me to shore, warning where rocks hide below the surface.

We jump a second time.

STATS
80.24 miles, max speed 22.5, ride time 7:40, avg speed 10.4, TOTAL 3115.42 (Avg 73 miles per day)

The jumpers. From left to right, Karina, me, Miguel, and Miguel Jr.

54

"And Sometimes my Stories are True."

Monday, August 1, 2016
Pelican Rapids to Chippewa County Park, Brandon, Minnesota

In the 1250-kilometer Paris-Brest-Paris (held every four years) the girl and her tandem partner rode with more than 2500 cyclists from around the world. They were strong and in sync as always.

They were the "Léopards Roses," the Pink Leopards, dubbed after their matching neon pink spotted tights. Reporters followed their progress and locals cheered from curbs. French riders surrounded them, pulling in a pack, and parted like a sea at the crest of a climb when the faster tandem screamed ahead.

"American men?" The French riders grimaced, shook their heads no. "American women?" Their eyes grinned. "Avec finesse," pumping right fists against their hearts.

Courage.

"The Muddy Moose," Miguel said last night when I asked him about a place to grab breakfast. It is nothing like muddy. The tidy cafe, tucked in a historic building on North Broadway, the main avenue through Pelican Rapids, is cozy with half-log walls full of local arts and crafts. A glass-faced wood-burning stove set into a cut-stone chimney dominates its center, but can't hide the back counter loaded with pies and pastries.

The right place that, like this town with its diversity and population of pelicans, captivates me. *I bid thee a fond farewell, Pelican Rapids.*

Back to the Midwest's rolling hills that snake from lake to lake. Amazing how quick the prairie loses its grip. The route today is south to southeast, and no surprise, head-on into a twenty miles per hour wind. My knees, which woke me in the night, no longer ache thanks to ibuprofen, but my legs are lead. Even so, I enjoy the coaster ride into Fergus Falls.

There—the school I passed—nice to drift down the curving hill instead of climbing it. Here—the left turn by a row of machine shops and warehouses looks different coming this way, as does a city park at the edge of a small lake. I don't remember the lake, but I remember the stately neighborhood lined with towering trees before the route spills onto the five lanes of busy SR 210. I slip past the Salvation Army Store where I searched for a smaller pair of shorts (never found a new pair, my chafing healed), and past the Subway where Tammy's gift card treated me to an air-conditioned meal.

I text Steve, "I'm in Fergus Falls, headed for the path. You guys doing ok? You missed a fun evening in Pelican Rapids. I met a family, and they were jumping off the suspension bridge into the water. They got me to jump too!"

"Awesome sauce. We are on Route 2, coming across. We have done a little over twenty…thirty-one from Fergus Falls."

"Great! Where do you think you'll end up today?" I doubt they'll catch me.

"Not sure. We are trying to get as far as we can, but the wind wants us to do something else."

"It is terrible. Sorry to have brought them on you. Be safe."

"You too. We were trying to catch up to you, but you're just too good for us."

"No, too bullheaded!" Didn't Mom used to say that about Dad? Five years before he died, I took his bike down from the garage rafters and sold it, because the only way he could dismount was to fall on the grass. He scoffed at his back surgeon's prognosis. "In a wheelchair before I'm eighty? I'll show him." He was eighty-seven before he needed four wheels.

Ah, two days away from traffic and steep hills. The Central Lake Trail skirts the Delagoon Recreation Area, Pebble Lake Golf Course, and the Hi-View State Wildlife Management Area. Everything is green, picturesque with lakes, grasslands, farms, woods, wetlands, and more pelicans.

Six miles down is Dalton, population 253. I pass a woman on a park bench, talking on her cell phone, a blue hybrid bicycle parked nearby. A wave hello. Beyond town, shade is a welcome relief. I recall a fox strutting ahead of me along this stretch.

Eight miles later I stop at a trailhead in Ashby to take my lunch. A few minutes later, the woman I passed rolls up. Sharon is riding from her family's cabin near Pelican Rapids while her husband and son play golf. "I'm meeting them in Brandon."

"That's where I plan to stay for the night. Care for company?"

"Sure."

Sharon towers over me on her upright bike. She is strong. I pick up my effort to stay next to her in this wind, good thing she does most of the talking. I learn she is a third-grade teacher, eleven years younger than me. She has two sons. Her oldest is entering college this fall, the other is sixteen. "His name is Brandon. I took a photo of him under the 'Welcome to Brandon' sign when he was a baby."

Sharon's dad died last July. A coincidence?

"My dad died last August. I think this trip is much about grieving him."

She nods. "I've had five other deaths this year."

Oh my. My heart burns. We cruise past Evansville, where I sat outside the corner store and chatted with a man who missed his dead wife. We pedal on and speak about grief and life's challenges. I am reminded how Lou and I shared conversation therapy like this when we trained together—my thoughts revealed while running, hers on the tandem.

By the time we arrive in Brandon, Sharon and I are hesitant to take our leave. We delay farewells with icy Gatorades in an air-conditioned gas station. Before we can take a sip, a weathered old man with bushy gray eyebrows and hair sprouting out of his ears saunters over. "How are you ladies today? I get bored sitting around. I come here to see what's happening."

Sharon notes a logo on his cap. "Are you a farmer?"

Her question unleashes a lifetime of stories. Myron, a local boy, was a bachelor farmer until he was thirty. "It was time to find me a suitable woman to marry. We had five kids and adopted a sixth." His eyes glisten. "I'm a story-teller, and sometimes my stories are true."

STATS
64.42 miles, max speed 34.5, ride time 5:40, avg speed 11.4, TOTAL 3179.84

Lou and me at the early morning start of Paris-Brest-Paris, 1987.

55

Absence Shows the Heart

Tuesday, August 2, 2016
Brandon to Bowlus, Minnesota

Twelve dollars. The prize the girl won, once, for placing in the top ten at a local criterium. Most times she got lapped.

Despite USA Cycling Hall of Famer Mike Walden and his elitist club, she bought a racing license and gave it a go on her own. She didn't care if anyone laughed at her Raleigh Super Course outfitted with plastic fenders and a rear touring rack. She was game to try anything that involved riding her bicycle. One fall she won first woman in the Michigan Cyclocross Championships. (No need to mention only one other woman raced.)

She had to admit it, though. "I'm not fast." Touring remained her passion. If leaving work for months at a time wasn't possible, she would condense her miles. How far could she ride in twenty-four hours? A lot, it turned out.

Years later, leaving a Bicycle Expo after giving tandem-roller riding demos, the girl, no longer even thirty, ran into Walden in the parking lot.

"Well, Patti, you've done all right for yourself."

Taste that sweetness. "Thanks, Mike."

Leaving camp this morning prompts the second time I reflect on my volunteer puppy counselor role with Leader Dogs for the Blind. The first was when I saw the Chippewa County Park sign on my way west. Absence should make the heart grow fonder, yes? Like how a month away from my high school boyfriend in 1974 showed me the truth of my heart? Conflict with the prison puppy initiative disheartens me. The pit of my stomach burns when I mull over resuming the monthly effort. Shouldn't I be excited? (That bothersome "should" again.)

What will I do when I get home?

The day heats into a typical Midwest sultriness like nowhere else. Air trapped under the canopy of trees smells musty, like a navy blue peacoat stored in a basement. No headwind this morning to cool my face. I tick off towns along the trail: Garfield, Alexandria, Nelson, Osakis. Osakis, where I startled the woman in the Community Center, where the Central Lakes Trail ends and the Lake Wobegon Trail begins.

West Union. Sauk Centre. I smile at the memory of the younger me, who sat in despair on that wood bench near Melrose, home of the cheery staff at Coborn's Store. I salute myself for taking that break and coming back stronger.

Melrose. Freeport, where I saw a recumbent rider scraping tar off his tires. No wet tar today, but my shins are filthy from dust kicked up by my tires.

The covered bridge into Holdingford extinguishes daylight when I enter, as I exit into the sun I suffer a second blindness. Like a beacon, Headley's Hardware Hank catches my now-I-can-see-again eyes.

The thing about hardware stores in towns like Holdingford (population 708) is they carry everything to fix anything. And sometimes everything else, from pet supplies to furniture, appliances to housewares, even bike parts. I don't need bike parts, but I'm sick of fixing the worn fabric on my seat with electrical tape. The horn of this cushy saddle looks obscene. Sure enough, a two-pack of micro-fiber kitchen towels and zip-locks will work for a reupholster job, with moisture wicking a bonus.

Leaving Hank's, I stop at the trailhead with flush toilets and the railcar where I paused riding west. Flowers cascade in stunning rainbows now from the painted tire pots. Seems a lifetime ago that I sat under the pavilion to FaceTime Mom, enthralled with sunshine lighting up fuzzies blowing from budding trees. Today I wash my shins in the sink.

It is a pleasant ride to Bowlus.

"Oh, I don't know," the young waitress says when I ask to talk to the owner of Jordie's Trailside Cafe. "Jordie's in back, working on her office." She motions to a closed door, behind which I hear a thump followed by a muffled curse.

"I stopped here on my way to Montana and Jordie let me set my tent up out back instead of at the park across the street."

The waitress shuffles, not excited about interrupting her boss.

"Never mind. Can I just pay the $10 to stay at the park?"
The girl's sorry face flips. "Sure!"
"And hey, can you make me a root beer float?"

Cooled off, campsite secured, I walk around the block to the grocery store. An old codger sits behind the counter at the cash machine. "Staying at the park are you?" Again my biker looks are obvious.

"Yep."

"Just to let you know, they might spray tonight." The grocer explains that every couple nights the city drives a truck around town, spraying overhead for mosquito control.

"What am I supposed to do?"

"Take cover."

"How will I know when they start?"

"Oh, you'll know. It'll sound like an airplane flying through town."

"Should I get inside the park bathrooms?"

"You'll be okay if you stay in your tent."

He is not reassuring.

And neither are the squadrons of mosquitoes on the attack. I retreat to my tent early. About 9:00 p.m. I wonder if I set up next to a landing strip. The grocer was right, the mosquito sprayer sounds like an airplane coming in for a landing. Despite the heat, I zip the rain fly shut and hope for the best. The pesticide reeks. I imagine a Stephen King mist seeping through the metal zipper of my tent, my lungs feel like they're in a vise.

To take my mind off, I compose an email to Mom. The anniversary of Dad's death is approaching. My eyes blur. "Just the mosquito spray," I tell myself, but I know better. My guess is that Mom's eyes will burn too when she reads my words.

My phone rings and buzzes while I type; I startle and fling the phone across the tent.

It's Steve. "We're in Sauk Centre. Wanted to tell you about the other night. A family in Fergus Falls took us in and rescued us. They own a restaurant and even though it was closed they opened it for us. We had chicken and all kinds of stuff. Awesome sauce! A bad day turned wonderful."

Sauk Centre is about thirty-five miles back. "We'll get up at six to try for the Bunkhouse. We're trying to catch you."

"If you get lucky with tailwinds, you might."

He makes me laugh when he tells me how he tricked Matt and Everett. "I told them you jumped off a bridge. You should have seen their faces!"

Ah yes, their faces. Everett/Matt, two sides of a coin: suburban boy/city boy; hybrid bike/racing bike; loves biking and started a cycling club at his school/is on this trip because the hiking trip he wanted to take didn't happen; organized gear/gear strapped on like Jed Clampett's jalopy; articulate and positive and eager to help/mutters about war during the rare times his ears are free of earbuds.

That night in Pelican Rapids, when I texted I was sorry I never got photos of us together, Steve sent a selfie of the three of them. His and Everett's smiles light up the evening. Even Matt managed a grin.

POATCARD FROM THE ROAD 8/2/16

Shade. Humidity. Shade. Mosquitoes. Shade. Water. Shade. Pretty easy to tell I'm back in the mid-west.

Did I mention shade?

Seventy-eight miles today on two of Minnesota's super-highways-just-for-bicycles, the Central Lakes and Lake Wobegon Trails. Wind? No issue, either way.

When my speed dropped I knew the grade was gaining. No worries. After a mile or two I'd find myself spinning out and gearing up. When a little extra effort brings your speed up to nineteen miles per hour, you don't mind pushing.

Somewhere along the Lake Wobegon Trail motivational signs appeared marking every half-mile, reminiscent of Burma Shave highway ads.

"Exercise is for life," "Exercise makes strong bones," "Commit to be fit," "Healthy life ahead," "Work out hard and don't look back," "Who do you exercise for?" "Exercise frees body, mind, and soul."

And my favorite: "Just one more mile."

STATS

77.79 miles, max speed 21.5, ride time 5:32, avg speed 14, TOTAL 3257.63

56

A Certain Freedom

Wednesday, August 3, 2016
Bowlus to the Biker's Bunkhouse in Dalbo, Minnesota

The first year the girl attended the Great Canadian Bicycle in Woodstock, Ontario, one scheduled event was a special showing of the iconic bicycling movie "Breaking Away." Wet cyclists packed the steamy theater after a rainy ride into town.

She cheered the Cutters on with everyone else, but a scene from the opening film stuck with her. In a documentary of Eddy Merckx, five-time winner of the Tour de France, a masseuse kneaded "The Cannibal's" granite calves into putty. Witnessing this transformation led to weekly massages as part of the girl's training. Ten years later, massage school.

It was fate, just in time for Rick's dying wife. "Liz asks for you every day," he said.

Powerful work. During one two-hour session Liz moaned, "I think when I die my energy will dissipate into the universe."

The girl agreed. "Isn't it that energy cannot be destroyed, only changed?"

As the end neared, a withdrawn Liz became edgy after a few minutes of massage, as if she were trying to leave her body and touch kept bringing her back.

No threat of extra protein in my pancakes this morning, the buzzing squadrons are grounded. My mind wanders to the guys behind me. My emotions see-saw. Overwhelmed at first by the crowd at the Honey Hub, I got excited for traveling companions. Disappointed to arrive at Enderlin alone, I was pleased when Steve and the boys showed up. I felt sad when the three stopped short of Pelican Rapids, but happy for their friendly encounter in Fergus Falls, and hopeful they would catch me.

Am I getting homesick?

The ride east is glorious. At the Mississippi, I notice a picnic area with boat access I hadn't noticed the other way. Passing the town of Royalton, it's a Holiday Station where I took refuge from the gale on Father's Day. There's the deserted house where I removed my fairing, where I sat on the grass for a rest and picked up that first tick.

Today I pass the Rum Shack without stopping. I do not miss my turn, I do not go two miles out of my way only to fight the wind two miles back. Today a side and slightly in-my-face wind doesn't bother me, and the hills roll more down than up. I grow stronger with every mile.

Due south before a turn east, the silhouette of a long, tall rider shimmers over the fields like a horseman apparition. A black hat on his head looks like an Indiana Jones fedora, but with a wide, wizardly brim. We intersect.

"You must be Patti." The lanky man of indeterminate age wears black, his tidy front and rear panniers are black, he towers over me like the metal ranch signs in Montana. My eyes rise. The joy from his scruffy face blinds me.

"Why yes, I am," I almost whisper.

"I've been at the Bunkhouse. Donn told me about you. I'm from the Twin Cities. My first tour, riding the Northern Tier from the east coast to the west."

"I'm heading home to Michigan."

He nods, as if he knows everything about me and says, "I hurry up to slow down. Rode through the UP in a day and a half so I can savor places I won't be near to explore in the future. I expect to spend some time in Glacier." He gazes over my head as if he can see the mountains from here.

We share a transcendental conversation about bicycle touring across the country, transcendental like my sparkling passage along the Blackfoot River to Ovando heading west. How pedaling pulls us beyond ourselves. How we are all mere droplets connected, star dust flowing from here to there and eternity. The adventure is more internal than physical, tracing progress on a map yet traveling beyond consciousness, hearing nothing but wind and the song in your head, smelling your past catch up to its future and believing in the now. Chasing dreams, collecting experiences like lightning bugs in a jar, savoring honey. That a certain freedom comes with needing nothing more than what you carry, of wanting to be nowhere but where you are, even if you learn you yearn for another person, still home is where we are and where we came from and where we are going.

On the pedaled road we stumble on peace. And pain exquisite. I am he and he is me, converging on an elevated plain. I have just met him, I have known him forever. I never even get his name.

I float the hills and cornfields. Materializing at Donn's pine-enfolded homestead feels like a homecoming. He strides out to greet me. "Welcome back."

"The three other riders I called about might not get here."

"That's fine. You'll have the place to yourself."

"Hey, did that Jesus-fellow ever come back?"

He gives a puzzled look. And laughs. "No...that was really something, wasn't it?"

It sure was.

Nostalgia saturates me, I wander the bunkhouse and copy several of the sayings nailed to the walls.

- Life is a beautiful ride.
- Life is not measured by the number of breaths we take, but by the moments that take our breath away.
- A quote by T. Roosevelt: "Believe you can and you're half-way there."
- And one from Tommy Lasorda: "The difference between the impossible and the possible is in a person's determination."

During FaceTime with Mom, she beams. "Joyce and Alice came over and took me to Chili's for lunch."

Alice is one of her older sisters. She lives in Florida but comes to Michigan to visit her kids. Joyce is Alice's oldest daughter.

"How is Alice?"

"She's doing good for ninety-two. But she has trouble figuring things out, like the bill at lunch and ordering."

No mention of my email. In four days, it will be one year since Dad died. Some while ago Mom told me he always brought her orchids. I never knew. *What if I send her an orchid?* I don't think too hard on it. If

I did, I'd worry that sending Dad's flower will add to her grief.

Instead, I text Steve. "You guys doing okay?"

A sad text back. "We are not going to make it. Sorry!"

"How far back are you? You might still think about a stop at the Bunkhouse. Donn Olson, who runs it, has info on good cycling roads. Because of him I'm changing my route tomorrow."

"We did more than seventy miles, but the boys missed the turn. We are southeast of St. Cloud at a town called Clearwater. Making dinner and doing laundry now."

"Yikes. Well, best of luck to you. It was fun meeting you."

"Same to you. We are in good shape, about sixty miles from Stillwater."

"Good deal! That's where I'll cross into Wisconsin tomorrow. I hope to get further before I stop so my ride to Eau Claire is doable on Friday."

My compadres are on their own.

POSTCARD FROM THE ROAD 8/3/16

"Ow!" If that wasn't a mosquito that stung my finger just as I grabbed my bar-end to shift gears, I was in trouble.

I shifted again and discovered it wasn't a mosquito. A few days ago I noticed a single strand of wire frayed at the shift cable. I twisted it off and lubed the mechanism, fully aware that it probably wouldn't end there.

Tentatively, I swept my fingers under the shifter. At least four more strands sprang out like frizzy hair in high humidity.

The rolling road the last twenty miles to Donn Olson's Biker's Bunkhouse forced me to keep shifting—ever so gently. As I pulled into his driveway, the sweet hum of an air conditioner promised a cool repair.

No telling how many more shifts before the extra-long cable would have snapped, leaving me in a "not so hill friendly" gear.

STATS

62.68 miles, max speed 26, ride time 4:51, avg speed 12.9, TOTAL 3320.31

57

On my own Route

Thursday, August 4, 2016
Dalbo, Minnesota to Willow River State Park in Burkhardt, Wisconsin

The middle-aged man's sweaty attitude blew into the bike store ahead of him. He swaggered to the service door stand where the girl pulled a crank off a $2000 full-suspension mountain bike covered in red mud.

"There a mechanic in today?" The man spread his feet as if preparing for a fight.

The girl set the crank tool and twelve-inch adjustable wrench on the bench behind her, wiped greasy hands on her shop apron, and stifled a laugh. In an instant, she decided.

"Sure. I'll go get him for you."

"Gotta kill some time." I phone Andy from the dry porch of the Bunkhouse. "It's raining pretty good." Mmm mmm, perked coffee made from real ground, belly full of farm fresh eggs from Donn's refrigerator, toast, and the pre-cooked bacon I picked up in Milaca yesterday. Thunder cracks.

"Looks like a big one on the radar," Andy confirms.

"Hoping it will blow through. Donn is funny. He and his wife drive the countryside to scout routes, recording mileages and notes about services. He says it's sixty miles to Stillwater. I'll cross the St. Croix River there into Wisconsin, shooting for Willow River State Park, not much further."

"You're making good progress."

"Should be at Karen's tomorrow."

"Gonna give you an 'ease into regular life,' meeting up with all your Leader Dog friends."

"Not sure I want regular life." The contrast between kindnesses I find on my traveling-road compared to the real world-scene is

Brobdingnagian, a colossal difference of a size straight out of *Gulliver's Travels*. Maybe it's just what plays in the media. Insanity rules, this presidential race is tiresome. "I can't believe Trump is a serious choice."

"What concerns me is the number of people who think the way Trump does."

"It must be fear. Fear of the 'other,' fear of change, of losing privilege."

Rain pelts around our talk of world trouble. No easy solutions. Again, I wonder if I would have had a different experience traveling as a person of color, something Joe and Jane and I discussed back in Montana. Does part of the interaction equation come with the level of receptiveness brought to the table? Don't we all affect each other? Take puppies, for example. A stand-offish, don't-touch-me puppy elicits a different response than a cuddly, please-just-pet-me puppy. Like the difference I felt from Jordie at the Trailside Cafe in June. On the phone she was curt, but in person she paused her rushing, smiled, and took me in with open arms. Would she have done the same the other day if I forced the waitress to interrupt her? Is the good karma I chance upon a result of the openness inherent in me? Do we reap what we expect?

We may reap what we expect, but that doesn't work with weather. Or wind.

I dawdle. Soon I will be home and this will be just another memory instead of an ongoing adventure. Will I miss the ride, the daily routine? Soon enough I'll find out.

Bidding adieu to the Bunkhouse, I wink at the spitting sprinkle. Andy's computer prediction is wrong again. The storm is long gone, replaced with sublime Midwest heat and humidity.

At Harris, I divert from the ACA Northern Tier to follow Donn's route, much of which includes the Sunrise Prairie and Hardwood Creek Trails. Ka-bump, ka-bump, ka-bump. Five miles on the broken up, narrow shoulder of SR 30 south to North Branch feels like fifteen. The bike path for an actual fifteen miles more isn't much better, both need serious repaving.

"There's a couple of miles of road construction before you pick up the new trail along the river, north of Stillwater. I think you can get through okay," Donn gave notice last night. "Send me an email on how it goes."

Traffic, steep hills, and bulldozed sand is not fun, but he is right, I manage. And then Hardwood Creek Trail. Woo hoo! Smooth as Andy's freshly shaven face. I brake and snake around mothers pushing baby strollers, walkers with leashed dogs, runners oblivious in Ear Pods. Gravity pulls a six-mile downhill run that dumps my racing heart onto a side road next to the St. Croix River.

Stillwater, a touristy town of almost 20,000 people, is seventy-three miles from the Biker's Bunkhouse. Don will get feedback tonight—bumpy narrow shoulder, jarring bike path, thirteen miles further than he said, but otherwise spot on.

Letting my heart calm, I consider. Take a break and turn right? The streets and sidewalks of Stillwater spill with vehicles and pedestrians. Or keep plugging left, into Wisconsin via the historic, two-lane Stillwater Lift Bridge?

I turn left.

Traffic on the narrow bridge bustles. I dare not ride, instead I squeeze my bike along a skinny sidewalk on the eastbound side. If another cyclist comes the other way we'll have a stand-off. The bridge empties its flow of vehicles into Wisconsin directly onto State Highway 35. I freeze. A few hundred yards ahead is my turn onto County Road (CR) E. Bunny-like, my eyes dart for options; fighting traffic that surges up a sharp curve (with no shoulder) is not my idea of fun.

There, a dirt path, overrun with thigh-scratching weeds, leads through the ditch to CR E, but my challenge is not over. *Drat these climbs out of river basins.* CR E looks like the Berlin Wall. Strike that. This hill is the Great Wall of China. *Why didn't I pack a grappling hook?* Head down, I push and pull and drag. *Good thing it's not as long as the Great Wall.* After a quarter-mile or so, I summit and rest my bike against a stop sign, dousing my head with warm water.

Eleven miles to Willow River Park rolls easy through farmland and forests, with low traffic. The park is not overrun, I score a site near the entrance, a short walk to flush toilets and showers.

Andy says, "I can tell you are ready to be home."

"What do you mean? Just because last night I said I want you to scrub me in the shower when I get home?"

He chuckles. "The tone of your voice. It's like you have less enthusiasm for the ride."

"I admit I'm pushing. It's my self-imposed deadlines: Wisconsin today, Karen's tomorrow. Not sure about getting to Manitowoc. Karen offered to drive me to the ferry, but I need to ride."

"Think you'll make it home in time to see Josh and Mandy?"

"I looked at the map. It's crazy, but once off the ferry I think I can be home in two days." Crazy, yes. A distance that a month ago would have taken me three days.

"I could come get you."

Of course. "I want to ride."

Mosquitoes buzz, crickets croon. What's that? "Coo, WOO! I. Love. You." I forgot the call of our Midwest-accent mourning dove. No more ear-battering THE BROOMstick, THE BROOMstick! Eyes closed, the sweet, mournful moaning carries my soul home.

STATS
83.96 miles, max speed 37.5, ride time 6:46, avg speed 12.4, TOTAL 3404.27

58

Between my World and This

Friday, August 5, 2016
Burkhardt to Eau Claire, Wisconsin

"I'm just going to check it out." The girl showed her husband the ad for a one-day Norfolk Southern Railway work fair. "A conductor on the railroad, now that's a bucket list thing," she thought. "Besides, the bike store can't keep supporting us, his ex and kids, and the previous owners."

The girl listened all morning to the trials of railroad work life: on-call with round-the-clock schedules and "It's not a question of if your train will kill a person, but when." Standing room only before lunch, the hall half-emptied when testing started after. The girl stayed.

Her mother's voice whispered from the past, "You are so competitive."

"Only with myself," the girl argued.

Now here she was, scratching out answers to pattern puzzles. "I will get this."

The sun sparkles wings of insects flitting over the roadway. The Wisconsin landscape of farm fields interrupted by stands of hardwoods, lakes, and rivers reminds me of Michigan. Hot to ride, I am determined to reach Karen's before five this afternoon.

The wind picks up with the temperature like a slow boil on a cook stove, evaporating thoughts through my helmet's vent holes. Dad's death anniversary is in two more days. Pedaling as therapy, the rhythmic effort sets my mind to the immediate.

Almost a tailwind. I dig into the hills. I am indomitable.

The road dips and crosses into another county. Here the asphalt is pristine, a foot or so of weed-free shoulder. My tires no longer rumble. They sing. They purr. "Smooth as a baby's butt." I grin out loud, not caring that the comment is an overused cliché. It is true.

A meadow rises to kiss a forest that cuts short the horizon. Surrounded by trees is a comfort, unlike gazing into the eternity of the prairie. My self is as connected to the earth as the turkey vulture above, circling thermals in a quiet waltz; my body sets to work on rolling the hill ahead. What goes down must return up. A true bicycling tenet.

Except, surprise, not this time.

The fresh paved road rolls down, down, and more down, fifteen glorious miles through the Knapp Hills all the way into Menomonie. Have I ever graced rubber on such a fine span of road as this? Like the year Dad and I won the Michigan National 24 Hour Challenge, my face aches. Nothing can wipe away my silly assed grin: not the teeming traffic in the town of 16,000 people, not the uncertainty about roads forward, not the heat, not anything.

I scream to a halt under a Visitor Center banner. Surely someone here can recommend a route to Eau Claire. A young man dressed in khakis and a polo shirt holds up a counter lined with brochures. "How can I help you today?"

"I'm riding my bike to Eau Claire and I'm thinking about taking County Road J. Do you have any maps or other suggestions for me?"

He sidles around the counter to a bookshelf, shuffles pamphlets, and comes up empty handed. "You know, I'll get someone else that'll be more familiar." He turns to a hallway. Three giggling women enter and waylay him.

"Where should we go for lunch?" one of them asks.

"Oh, we've got lots of great places." He draws them to a city map at a second counter.

Meanwhile, grateful for the air conditioning, I browse.

Eventually the man remembers me. "I'm sorry. Let me get my boss." He disappears. A business-dressed woman appears a moment later. She digs in a cupboard and retrieves a local map.

"You're trying to get to Eau Claire?"

"Yes, how is County Road J for bikes?"

She spreads open the map and points. "You might do better on County Road E instead." Funny how Wisconsin names their county highways with letters. The road she recommends is not the CR E from yesterday.

This CR E turns southeast about five miles past CR J, a tad shorter to where I'm headed.

"Where are you coming from?" With my answer, questions sluice like she's panning for gold. "Can I take your picture?"

"Of course!"

High heels click back to her office and return with a point-and-shoot camera. "I want one with you and your bike." The sight of my recumbent leads to another barrage.

I am overrun with her enthusiasm. "I tell you, these last fifteen miles into town on 12 was the absolute best stretch of road my entire trip!"

"You mean coming in from Knapp Hills?"

"Yes. A beautiful downhill run on smooth pavement. I wish all roads were like that one."

Ms. High Heels's advice leads me to this: LOCAL TRAFFIC ONLY.

The towering construction sign anchored with sandbags screams that the road is closed in five and a half miles. I am inclined to ignore the warning. On only two occasions has this ever proved a poor choice. Once, a bridge was out. I took off shoes and socks and carried my bike across the narrow creek. The second time was with Andy on our tour to Green Bay, Wisconsin.

After a rocky roll across Lake Michigan on the ferry (thank you Dramamine), we hoped the fewer-than-fifty miles pedal to Andy's son's house would go smoother. Too bad my hubby is married to the Headwind Queen. It was a brutal ride.

And then came the "road closed" sign. Andy deflated.

"Oh, come on. Bikes can always get through," I cheered. "We'll be fine."

Against his wishes, he followed.

Ten miles later the road ended. Literally. We stopped on the brink of a deep cliff above a wide river. No way to ford this one. Andy was not a happy camper adding twenty miles to our day, even with a tailwind assist on our retreat.

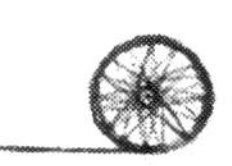

Today I go for it, even if lack of traffic portends a risk. Wind carries me like a prairie schooner; smooth asphalt, clean shoulder, and a gentle grade bring me to the actual "road closed" in no time. I check my map.

A motorcyclist pulls up and cuts his engine. "Need help?"

"Should I take a different road? Not sure if the bridge ahead is out." And not sure how far I want to fight a headwind if I need to turn back.

"It's clear. I'm going that way myself. You'll be okay. They're working on the shoulder, haven't started on the bridge yet."

He deserves a hug. Except for a short section of dirt where backhoes are digging ditches, I sail clear to the end of the construction zone, with no traffic. A field of sunflowers applauds my arrival.

Yesterday Karen texted detailed directions of the last hilly miles that I follow to her house. I coast to a stop at a green gate. A paved driveway curves right a football field length to the house, on the left is a horse barn, the right a wire-fenced garden. The yard is lush grass and shade trees. Several puppy counselors are unloading their cars. Karen is hosting these dedicated volunteers for the weekend.

"She's here!" someone yells. "Patti! You're like having a rock star come to our camp."

I dance a boogie. The time? A bit past 3:00 p.m.

Karen leads me to a pop-up tepee camper like the one Verna and her daughter had at the Big Sky Camp & RV Park in Miles City, Montana. "All my spare rooms are full. I didn't think you'd want to stay inside anyway, so I got this ready for you."

"I don't mind setting up my tent."

"Nonsense. You'll have this to yourself. There's plenty of time to clean up and get laundry started before we head to the training facility at five. It should take us about an hour, then we'll come back for dinner. Here are towels and I'll wrangle up clothes for you."

Before stepping into the bliss that only those who have lived outside for days on end can appreciate (never mind that my tent could fit in her tile and stone walk-in shower), I step on Karen's digital scale—117.75 pounds. More than ten pounds lost since June 9. No surprise the way my clothes keep growing.

Eyes closed, I stretch, and let the steamy water release me.

Not a camp towel, a real, fluffy bath towel invigorates my skin, sun-kissed and unexposed alike. For kicks, I step back on the scale—117.5. Could I have washed off a quarter-pound of sweat and grime? Maybe.

Karen left me underwear, shorts, t-shirts, and a bra. I slip on the underwear and pick the best fitting khaki shorts and a teal t-shirt. I set the bra aside.

Now I'm hungry. Wait. Since leaving Missoula I've been ravenous. Lucky for me, fresh fruit, cheese and crackers, veggies, and various dips mound Karen's kitchen island. While everyone else readies supplies and discusses plans for tomorrow's puppy camp, I graze, transposed from the free-range plains in eastern Montana.

At the facility, the women are fierce in preparations. I help how I can, dragging in crates and platforms, placing chairs and tables where they tell me. I am glad I brought a water bottle and a Kind bar. By 7:00 p.m., when the counselors run through the agenda, I wish I had grabbed my food bag too. I drool over the thought of its contents: cheddar cheese, wheat tortillas, a banana and apple, even leftover Cheeze-Its.

It is 8:00 p.m. before we leave.

Barbecue wafts as we pull into Karen's driveway, her husband Marty tends a grill. "Everything else is inside and ready. I just need to pull the veggies off."

Not just veggies. A feast awaits with a swimming pool of marinated red, green, and yellow peppers, mushrooms, onion, zucchinis, and who knows what all. I sip my first glass of wine since dinner with Susan in Helena. In fact, I have two. I barely hear the conversation around the table I am so busy foraging pork fajitas and guacamole. I don't care if I appear to be a Neanderthal caught between my world and this, I am among friends.

First things first.

STATS
72.91 miles, max speed 38.5, ride time 5:10, avg speed 14.5! TOTAL 3477.18

Thumbs up for beating the detour.

59

Halfway House

Saturday, August 6, 2016
Eau Claire, Wisconsin, Karen's

Two weeks at the Norfolk Southern Railway headquarters in Atlanta was like fantasy camp. The girl threw switches, started diesel generators, set air brakes, connected knuckles, and learned signals. Pre-sunrise calisthenics and stretching exercises in formation reminded her of ancient tribal rituals.

The only woman in class, she was old enough to be the other students' mother. When one of them over tightened the wheel at a handbrake training station, she couldn't break it free. "More than one way to skin a cat." She looked for a pry bar.

The boys laughed. The instructor pushed them aside to grapple with the wheel. "Move on," he commanded, and closed the station.

Back at her home yard, the girl realized little had changed since her first government-mandated, must-hire-women job almost twenty years ago. "You stay inside." A co-worker stepped out to build trains without her. Before the door slammed, a mumble. "Damn pissers."

Three days later she gave notice.

At the sound of sprinklers, I leap and almost fall out of the bed. Water jets a pattern across the side of Karen's Taj Mahal camper. I rub my eyes. "Oh yea, I'm not in my tent…my bike!" Timing my advance, I rescue my steed and race it to safety.

"Oh, shoot." Karen offers a bowl brimming with homemade granola, bananas, blueberries, and strawberries. "I could've turned off the timer."

"No harm done," I slurp.

The puppy counselors are all business getting organized. I will not make yesterday's mistake. Food bag? Check. Food? Check. I toss in my notebook and a water bottle and grab my Nikon.

As Andy predicted, puppy camp is like a halfway house to normal life off the road. Meeting several puppy raisers in person that I've only known through Facebook is a pleasant diversion. The group of thirty-six volunteers (and eighteen puppies) are pleased to see me too, and not only for my road stories. They spot my camera. I've been shooting photos of Future Leader Dog puppies for a long time and Leader Dog begs to use my best shots in promotional materials. I don't mind. Being shielded behind the lens keeps me in that halfway place. For a while, anyway, until Deb Donnelly solicits me for a demonstration.

"Patti is my dog." Deb hands me the other end of her leash. "Let's see how she does walking past all of you. You're a big distraction for her."

I play along. I've seen Deb work this help-the-puppy-make-a-good-decision exercise and have done it with Tammy at the prisons. I wander toward a fuzzy golden retriever lying on a mat next to its woman handler. *Wait. What's in her hand? A bagel?* That I hold the leash is gone from my mind. I step forward, I demand to know where she got the bagel and if there is cream cheese. Deb holds tight. My hand arrested, I strain to reach the woman. Nope. Not going anywhere. I glance at Deb, annoyed.

"Yes!" She gives me a Halloween-sized pack of M&M's. Forget the bagel, I move closer to Deb. She resumes walking. Everyone laughs, even me. I completely got into character. I could not have "acted" it any better.

The day is jam-packed with hands-on learning. Raisers use platforms to practice treat delivery and encourage proper position for "heel." Ground tethers are a tool to help puppies ignore distractions. To reinforce a loose leash, handlers treat at every orange cone staged throughout the training space.

The cones lead teams to simulated, real-life training stations. Funnel in line past a ticket booth to a theater seat (tempted by popcorn litter). Work an obedience rally course (signs instruct to "put your puppy in a sit and walk around it," "turn left," "have your puppy lie down for ten seconds," etc.). Drag luggage through airport security (remove shoes and empty pockets, get scanned and patted), then sit for a short flight served by attendants with beverage carts. Visit a veterinarian, including a wait with other dogs, enter the exam room, weigh the puppy and place on a grooming table.

By the end of the afternoon I want to curl up on the floor with all the snoring puppies dreaming of dinner.

Ah, dinner. Despite my gallant efforts last night there are enough leftover fajitas to support another feast. Karen and I find ourselves alone, shucking cobs of corn on her front patio. My phone rings a FaceTime request.

Mom sits at her usual place in Dad's old blue chair. "Where are you?"

"I'm still at my friend Karen's. We had a Leader Dog puppy camp today."

"Oh yes, I forgot." She can't hold back a sob. "I miss him so much."

I tear up. "I know, Mom."

She can't talk. I wait. With her, and not with her.

"I gotta go. At least I know you are safe tonight." Click.

I break down. "We lost my dad a year ago tomorrow." Now Karen waits with me. And listens. "I'm sending an orchid to her, I guess Dad always gave her orchids, I hope it doesn't make her feel worse."

We take a long time shucking corn.

STATS 0 miles

60

Mission Accomplished

Sunday, August 7, 2016
Eau Claire, Wisconsin, Karen's

This conversation, en route to the hospital for another shot to ease her father's back pain.

Mom: "I don't want my body displayed in a casket. It'll be okay for you guys to view me, but I don't want to be preserved."

Dad: "We want to be cremated. Mix our ashes and take them up to Marquette to sprinkle them in the lake. She left her home in Marquette to be with me. Only fair that I should leave my home to be with her there. We'd be together."

Even if they were pushing eighty, and she was no longer a girl, this end-of-life talk was hard to take. The girl could not stop her tearing eyes.

Mom: "You cry a lot because I cried a lot when I was pregnant for you."

Girl: "I remember the stories. Dad went on strike before I was born. He worked whatever jobs he could find. What were they again?"

Dad: "I was a waiter at the Belkan Club, sold Fuller Brushes door to door, and chopped ice for people after an ice storm."

Mom: "And I sold my urine to a drug company. I collected it in a jug that got picked up once a week."

Text from Steve. "Good morning, Patti, we did about a 100 yesterday and we are in Strum, just south of Eau Claire. We rode US 10 all day, it was a good road. Not much traffic and a decent shoulder. Are you still with your friends? We hope to reach the ferry by Tuesday."

Strum is less than twenty miles from here. I'm torn. Karen's husband is leaving for an extended fishing trip, the counselors have packed and hugged goodbye. I hoped to spend the day with Karen, exploring our growing bond. I never thought I'd see the boys again, but if I leave today, I can catch them.

"Stay," Karen says. "I'll take you to wherever they are tomorrow."

Text back. "I am still at my friend's house. She's offered to drive me to meet you tomorrow. Where are you headed? We can ride to the ferry together!"

"Awesome sauce, the boys will be glad to see you. I'll keep you posted."

Karen wants to show off her prairie, part of the eighty acres she and Marty own and are coaxing back to native grassland. We tuck long pants into our socks and spray with DEET. No more ticks for me. Karen's six-year-old black lab, River, circles in anticipation. Luna, a nine-year-old, 110-pound Anatolian Shepherd belonging to her eldest daughter, lifts her head out of a bucket. She's chewing something.

"Oh no," Karen says. "I left the corn cobs out."

"You know what that means."

Karen nods. "Time for hydrogen peroxide." She hustles both dogs into separate, custom-built kennels in the garage and retrieves the peroxide. She's brilliant, she put it in a turkey baster slathered with peanut butter.

Karen holds Luna while I present the baster. The dog sets to licking and I squirt the peroxide down her throat. Now we wait. It doesn't take long. She barfs several piles of corncob mixed with undigested dog food.

Reminds me of last winter's debacle with Gus.

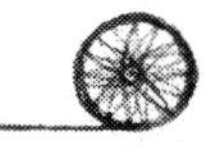

One Wednesday morning, before meeting Tammy on I-75 for our monthly prison trip, Gus ran off. He returned with a raw, dinosaur-sized bone. Wouldn't come to me, wouldn't drop it. Only when he ingested the entire thing did he come to me, licking his lips.

In a scene worthy of top prize in America's Funniest Home Videos, Gus bucked me into the snow when I pried his mouth open for the hydrogen peroxide. "Andy, bring the baster!" Should have used Karen's peanut butter trick. Andy couldn't hold him. Gus and I were both soaked with peroxide. Did any get in him?

Late, I couldn't wait around to see if the peroxide worked. Poor Andy got stuck with everything. An hour later he called. "He barfed."

Well, Gus barfed. And barfed and barfed, all day, all over the house. He stopped pooping so Andy took him to the vet. Six hundred dollars later, x-rays showed six inches of bone in his colon. The vet prescribed drugs to ease his stomach and lube the passing. "If he doesn't poop in the next few days," he said, "he'll need surgery." A $3000 surgery.

When I got home Friday night, still no pooping. Saturday we took him back to the vet to avoid a Sunday emergency. Before we loaded him into the car, he pooped. Andy found a small section of bone turned black. We took him to the vet anyway.

"It's coming." The vet slid his gloved finger from Gus's rectum. "I can x-ray but I don't think we need to. You'll just have to check to make sure he passes it all."

The experience made our much-debated decision easy. A $3000 surgery or a $1500 fence? Before I left on this trip, I helped dig post-holes.

Disaster averted, Karen and I stroll through grasses taller than me sprinkled with wildflowers. We amble under the shade of ancient oaks, bending to clear broken limbs from an overgrown trail. Karen talks and talks while we walk and walk. My turn to listen. She tells me of losing everything in a devastating house fire a few years ago, the heartbreak of rebuilding, her mother's death in May 2015, the strained relationship with her father, the indefinite Luna dog stay while they help their daughter her through a crisis.

Laurie Anderson's voice rises from the base of my skull. Her lyrics about lifting the fallen, holding the broken resonate. She sings of not knowing our origin, not knowing what we are. What we are is connected. Life is a beautiful wood. Life is loss. Sharing each with grace and empathy lifts us. I am meant to be here. Now.

Another rendition of leftovers leaves me with a renewed appreciation for refrigeration. We settle in Karen's sunroom with a grand view of the property. She fidgets at one end of a couch. "I want to ask you about the prison program."

Karen is a long-time puppy raiser for Leader Dogs for the Blind and a well-respected counselor. More than 400 raisers, throughout twenty-two states and Canada, use training materials she developed. Last March, Leader Dog's new Coordinator of Prison Puppies roped Karen into helping set up a group in a prison near Eau Claire. "The intention was that I'd give up my outside raisers to become the prison counselor, but I want a plan in place for my own raisers before committing. They've done nothing about that. I'm not feeling good about the support from Leader Dog. You've been involved for a while. What's your take on things?"

"Run away as fast as you can." I surprise myself.

Yes, the prison initiative is a fabulous win/win/win for Leader Dog, the puppies, and the inmate raisers. Yes, I'm proud to be part of it. But. The program grew too fast without a clear set of policies and procedures in place, with too much responsibility on the shoulders of under-appreciated volunteers. Until this moment, I hadn't realized how frustrated I've been in advocating for my prison.

"Your gut feelings aren't overblown," I add.

"I just don't want to get into the thick of things and be sorry. It'll be hard to back away once I take it on." Karen likes things organized with clear expectations.

I confirm her fears. "I don't know how long I can keep going myself. I haven't missed it at all. You've got a lot on your plate. If you do it, your stress will only go up. You need to think of yourself."

"And my raisers. I won't leave them without a replacement."

"If you take on the prison job without one, you'll end up doing both."

Karen sags into the couch. Luna creeps up to snuggle.

These two off-days have been a slow reintroduction to my other life, spiced with the deep wonderment of a new friendship. I have a sense of mission accomplished.

Later, FaceTime with Mom is less fraught. She gleams. "I love the orchid. It is beautiful. I posted a picture on Facebook."

A text from Steve bings in. "We are in Granton for the night. We'll stay in the city park and eat at Tommy's Hilltop Tavern. Didn't get too far today but we'll get up and going early tomorrow."

Time to get back on the road. And home. *Miss you Dad.*

STATS 0 miles

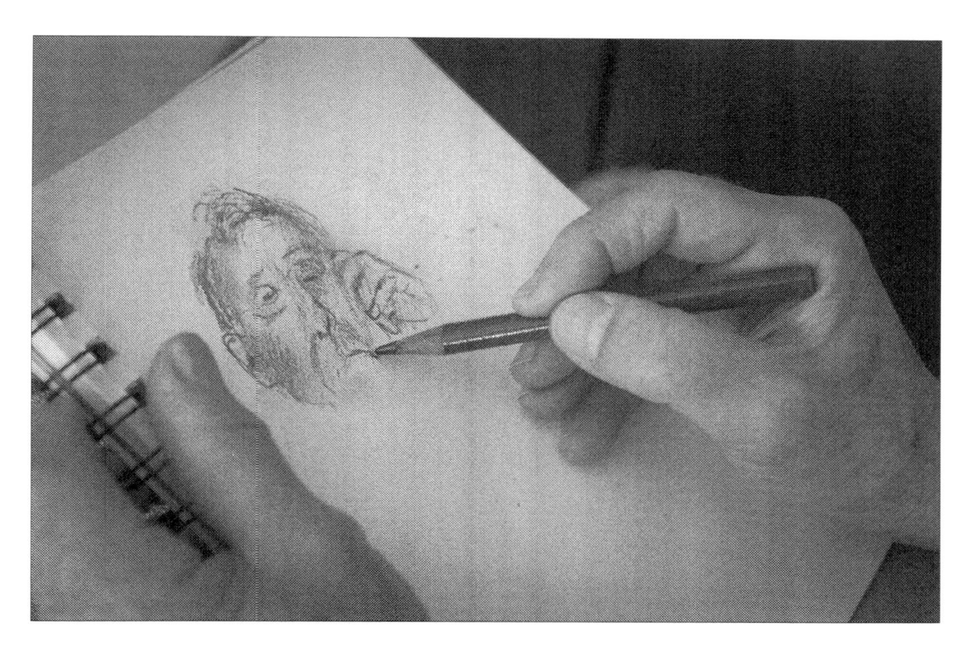

Cathy sketches Dad a few months before his death.

61

Once Again Comrades

Monday, August 8, 2016
Eau Claire to Emily Lake Park, Amherst, Wisconsin

"Your father would get up with me when I had to feed the babies. I told him, go back to sleep, you have work in the morning. But no, he stayed right with me."

his hands
deep waters covered in chill skin, like ice
crystallizing on a calm pond in November

but it is only July

he drapes
blue-veined fingers over mine, resigned,
he cannot cut his own nails any longer

by August we say goodbye

With my stripped-down Tour Easy hanging off the back of Karen's SUV, we drive in less than two hours what would have taken me a day to ride.

"This place is a favorite." Karen pulls into a lone restaurant and cheese shop off US 10 in the middle of nowhere. "Their specialty is brick-oven fired pizzas and they have local music on the weekends." I make it my treat. It is the most she will let me do.

"Our kitchen isn't open yet," the young woman behind the counter says. We are the only customers.

"We just want a simple sandwich. I'm taking my friend here to meet up with some other riders. They've been riding their bikes across the country."

"Okay, I suppose I can whip something up for you."

We order coffee and settle at a cozy table. Not even through our first cup and the woman brings crispy grilled cheese sandwiches on artisan bread and a pile of house-made chips. I wolf everything. Karen saves half of hers to take home.

Like-minded in our encounter, we linger to savor our connection.

My phone bings. "It's Steve. They're at a bike shop in Marshfield. You can drop me at SR 80 and I'll ride to catch them."

"Find out where," Karen says. "I'll take you to them. I want to meet them."

Karen hangs with us while I unload my bike, reinstall the fairing (taking better care of the mounting screws this time), and reload gear. Matt's bike needs work again, this time a new bottom bracket. It's been a long time since I've been in a shop. A counter fronts the open service area at the back door, and three work stands, with bikes in various stages of repair, coordinate with the usual tools hung over a long bench. Matt's bike is in the center stand. The setup reminds me of our shop's side-door stand, where I often wrenched on bikes and helped customers.

Having a friend from my other life mingle with new friends from my road life is disorienting. I stand between worlds, at once belonging in both and not.

"Hey, we're going to Subway," Steve says. "Do you want a sandwich? My treat."

Karen begs off. Me? No thoughts of the grilled cheese and the extra two pounds on the scale after this morning's shower. "Sure. Six-inch tuna fish with lettuce and tomato on nine-grain bread." When Everett returns with the loot, Karen bear-hugs me goodbye.

Lunch on wood stools in the showroom, surrounded by racks of colorful cycling jerseys, fresh-out-of-the-box bikes, and the smell of rubber is nostalgic. I don't understand it, but I am delighted to be back with my comrades, relishing tales of their last few days. The banter between Everett and Matt is as dear as the anthem of a coasting freewheel.

"Our plan is to follow US 10 all the way to Manitowoc," Steve says.

"Bikes aren't allowed on expressways in Wisconsin," I counter.

"It's a divided highway for a while here in Marshfield. We'll figure it out as we go."

After twenty miles of gradual grades on a wide, debris-filled shoulder, time to figure it out. Luckily, a local road roughly parallels the freeway. And we roughly stay together (sorry boys, the Headwind Queen is back). We negotiate broken sidewalks through two small towns with main streets torn up for reconstruction. Hills steepen. Heat steams humidity.

And still I smile.

By late afternoon we cruise under the shade of mature trees in Stevens Point, population 26,000. At its eastern edge we stop to splurge for dinner at Culver's. I am familiar with the almost, but not quite, fast-food hamburger chain, having indulged at the Culver's in Gaylord, Michigan with Tammy during our UP prison trips.

"That looks good," Steve says as I dive into cranberry bacon bleu salad with grilled chicken and sides of green beans and mashed potatoes. He heads back in to order the beans and potatoes to complement his butter burger.

When we are ready to go, Matt is AWOL. "How can he be so selfish?" Steve fumes. "He always makes us wait."

"I doubt he is even conscious about it."

"What do you mean?"

"Don't you wonder about his life? He doesn't seem happy. Sometimes I think when kids feel out of control of their lives, they try to control anything they can. When my husband's kids were young, we never got definite schedules until the last minute. I wondered if it was their way of trying to be in charge of something after their parent's divorce."

"I think he's just lazy. No matter how much I nag him he won't get it together."

"So nagging doesn't work. Maybe try something else. People aren't much different from puppies. If you reinforce the behavior you want and ignore the behavior you don't like, they'll start offering the behavior you want. Surely you can find one thing he's doing the way you want."

"Oh, I've tried that. Didn't work."

Steve's dismissal reminds me of puppy raisers who ask for tips to solve a behavior problem, but insist they've tried whatever I suggest, to no avail. They don't consider it might be bad timing with reinforcement or not paying enough attention to catch good behavior. Why are we humans so quick to bark orders and demand compliance?

Eventually a sullen Matt reappears. Off we go.

CR HH weaves among forested subdivisions out of Stevens Point. And straight up hills. At least they are short. My downhill speed rolls me ahead of the others. I'm not used to riding so late, it is after 8:00 p.m. by the time I reach Emily Lake Park. Fading daylight is keen under canopies of hardwoods and pines.

FaceTime from Everett. "Matt has a flat." He turns his phone to show Steve bending over Matt's rear wheel.

"Need help? Or should I get us a site?"

Steve looks up. "Get us a site. Thanks."

I text directions to a secluded spot high over the lake. The three riders wheel in before needing headlamps to set up their tent. They get to work in silence.

STATS
50.17 miles, max speed 25, ride time 4:30, avg speed 10.7, TOTAL 3572.35

From left to right, Matt, Everett, Steve, and me in front of Karen's car. Photo by Karen Voss.

62

"Hey, Everybody Poops!"

Tuesday, August 9, 2016
Amherst to Brillion, Wisconsin

Highway Deaths

Here, white crosses shock the landscape
like carcass ribs glinting in grass
fun squelched in a brake squeal
another self
snuffed.

There, numbers on a billboard blinked
like June fireflies, 32 that week
17 the week before
982 by year's
end.

Dear Dad died in bed between
numbers 585 and 586,
nothing blinked and no white cross
raised to remember him
bye.

"Matt's tire is flat again," Steve greets me in the morning.

"Let me check it out."

"Get over here and watch what she does," Steve barks as I ruffle for tire irons. Matt plucks out his earbuds and saunters over.

"First, I slip one iron under the bead, pull it over the rim, and secure the hook end to a spoke. These Trek rims are always tight. At our store, we had a special tool to get them off. Then, the second iron goes here, close to the first one."

"Why do they call those tire irons?" Matt asks.

"In the old days they were metal. I've snapped more than one of these plastic gizmos. Slip the third iron around the rim to free one side, but leave it on, only remove the tube. That way we can check the tire once we find the hole in the tube."

A long crease on the side of the tube hisses when I pump air into it. I locate the corresponding spot on the tire sidewall.

"This isn't good." I peel the tire off the rim and Matt leans in for a look. "Here, see that wire? The tire pulled away from the bead. It's toast. Do you have a spare tire?"

He retrieves a worn, foldable tire buried in his pack. "We took this off a long time ago. I was getting lots of flats." *It will have to do.*

"Tell you what Matt, let me carry your tent today. It might help to take some weight off that tire."

No argument. Now my load looks like the Leaning Tower of Pisa.

So much for an early start, by 10:00 a.m. we are only three miles down the road. Might be a long day. Steve wants to get as close as we can to Manitowoc tonight to make the ferry by noon tomorrow. I hang with the boys as long as I can, but undulating terrain rolls me ahead.

A prehistoric chatter from a crop field (*are those potatoes, Lou?*) catches my attention. Four sandhill cranes tumble and flap like awkward pre-teens on a basketball court. They rise as one. How can creatures so uncoordinated and gangly on the ground become so synchronized in the air? A couple of wing flaps soar them over power lines. In silence, they float a gentle hypotenuse into the southwest wind, gliding to a few feet above the ground. Their graceful landing under shade trees lining the field carries my breath to peace. My mind meanders with the road, between farm fields and forest, this life and the life to which I'm about to return.

The boys catch me at a tee in the road, where a hard-luck bar squats in a gravel parking lot. "Go ahead." I nod our right turn. "I need to run in there to use the bathroom."

Inside I stop to adjust my eyes. A middle-aged couple sit on high stools at a u-shaped bar, a woman about the same age tends them. The man shakes an oversized pair of dice in two hands and clatters them across the counter.

"Can I help you?" The bartender wipes her hands.

"Can I use your restroom?"

"Sure thing, it's back there." She flips her towel toward the far side of the room. I clomp across the dusty wood floor to the one-hole bathroom only to find the toilet plugged. I flush. Seems okay. After my relief, nothing happens when I flush. There's a plunger behind the tank. No luck, the tank doesn't fill up. I remove the top to check the chain but nothing looks amiss. I fret over this awkward situation before coming out to confess my regrets.

Before I can say a word the bartender says, "You ok?"

"Something is wrong with the toilet. I tried plunging, but it won't go down."

"Did you flush more than once?"

"Yes." *Can my head hang any further?*

"It won't fill up if you flush it too soon."

"I'm sorry." She can't see me blushing, she must read something in my tone of voice.

"Hey, everybody poops, don't worry about it!"

The couple at the bar toss off my embarrassment as if it happens all the time. They ask about my ride. "Michigan to Montana! Are you crazy?" the man says. "How far are you riding today?"

"We've got thirty miles in so far, we hope to get close to Manitowoc tonight. We're catching the ferry tomorrow to Michigan. I'm almost home!"

"Thirty miles! I couldn't ride thirty miles in my life."

That he probably could might surprise him.

Time wasted toilet wrestling makes me work to catch the boys. They've stopped at a gas station convenience store.

"I'm having a bathroom problem myself," Steve confides. "It's the heat." When we hit the road, Steve takes up the rear. Everett and Matt pedal hard to hang with me. At the top of a long hill they roll even.

"We had to catch you on the hill because we knew as soon as you crested you'd be gone," Everett says.

A little less push on the downhill and he and Matt stay right on my wheel. I am proud of their determination.

Twenty-plus miles later, Appleton, with a population of almost 73,000 people, confronts us with rush hour, on gnarly roads. Steve defers our route finding to my iPhone. A five-lane road should drop us through to the south end of town.

Years of experience riding in urban traffic and still my hands cramp gripping the handlebars. I keep one eye on my mirror—when I notice Steve direct the boys onto the sidewalk, I follow suit. Too many cars, traffic lights about every half mile, and delivery trucks turning in front of us make for a slow and arduous passage. We sweat like leaky buckets.

"I dropped my tubes," Matt yells after negotiating a tricky left-hand turn. The tubes fell out of his backpack that hangs off the wobbly pile strapped to his rear rack. The west hanging sun blinds us. Yep, there they are. A turning car runs over one, bouncing it a few feet further.

Steve grabs Matt's bike. "Go get them." Matt hangs his head, much like I did in the bar, and does as ordered.

We discuss options. "If we keep south here, we can stay at the High Cliff State Park," I say.

"I think we can get to Brillion." Steve points it out on the map some twenty miles further. "Less miles to Manitowoc in the morning. And we can get back on US 10 the whole way."

I check my phone. "I don't see a place to camp."

"I'm sure we can find a city park."

"I know out west you can camp at city parks, but I'm not sure about Wisconsin. I don't think you can in Michigan."

"We'll be fine."

I'm not so sure but there is confidence in numbers. And it does make sense to push now for a shorter day tomorrow. We make it to Brillion by 7:00 p.m. Everett and Matt pick up their pace when Steve points us to a Dairy Queen at the top of a long hill into town.

No sign of a park, we stop at Tadych's Econo Foods. Once again I drool down the aisles like a Labrador in the pet food section. Steve loads a cart with veggies and rice and cereal for tomorrow's breakfast; I grab a ready-made chef's salad and a chicken breast so huge it must have been pumped with hormones. I don't even care.

"Do you know where the city park is?" Steve asks the cashier.

"Well, there's Peter's Park south of town and Horn Park just a couple blocks that way." She points east.

A customer standing in line behind us adds, "Horn Park is your best bet. Lots of trees."

Streetlights are on by the time we exit the store. *Time to get home*, a childhood memory whispers. Tonight, home is Horn Park, around the corner from a senior care facility. The customer was right about the trees. We follow Steve along a shadowed paved path like a fellowship of hobbits behind Gandalf, past a restroom building, and up a short hill to a lighted pavilion.

Perfect.

We execute chores with precision. Steve's veggie stir-fry smells divine and I don't hold back when he offers me a taste; I slice and sauté the chicken breast to share back. Matt watches me secure my iPhone with a bungee cord to one of the square metal poles in the pavilion, so the charger reaches an elevated outlet. He does the same with his.

Motion sensors flip decorative lights on as I walk to the restroom. If I get up in the night, I hope the lights don't wake the others. No worries. The three snorers in the next tent almost drown out the midnight shift in the factory across the way. All night the roar is like sleeping next to Niagara Falls, beeping forklifts and banging presses keep time.

STATS
84.11 miles, max speed 29.5, ride time 6:39, avg speed 12.7, TOTAL 3611.75

63

My Previous me Doppelgänger

Wednesday, August 10, 2016
Brillion, Wisconsin to Manitowoc ferry to Ludington, Michigan

Wide Load Overtaking

One
eye
mirror
peering back,
get out of the track.
The past—let it pass, let it pass.

"I stayed right with you." Matt grins, dripping up to me as I sit feet down at the turn onto CR P leading into Manitowoc. Topping eighty degrees before 7:30 a.m., the asphalt of US 10 steams like Yellowstone fumaroles.

"You're doing great."

Steve seems not so great, almost out of sight behind a soon-to-arrive Everett.

One of the tough things about bringing up the rear is the poor chance of getting much rest when you catch the pack. Steve barely sets his foot on the ground when the boys and I lift our feet to take the turn. Matt and I pull ahead. I ease the pace.

In town, my tires shimmy on the steel grates of a drawbridge over the Manitowoc River. Concerned about Matt's skinny tires, I watch in my mirror. He follows straight on and makes it fine.

"What were you thinking Matt?" Steve rides up. We're stopped at a red light. "You know you're supposed to ride on the sidewalk over bridges like that."

I jump in. "Hey, it's my fault. He was following me."

Steve isn't happy. "We'll stop at that Subway and get sandwiches for the ferry." We walk our bikes across when the light changes. I wait outside, my food bag is full.

As they secure their lunches Matt asks Steve, "Is it okay if I listen to music when we're on the ferry?"

"No."

"Why not?"

"Really? You want to get into this now?"

"I'm just asking."

There must be a history between him and Steve with the music. Even so I feel a pang for the young man. We follow signs to the ferry dock in silence.

Steve cools off in the air-conditioned ticket building while we wait to board. *Thank you, my dear friend Debbie, for your generous gift of my ticket—at last you found something to help me on my way.* Outside, I steal a smile from Matt when I take a selfie of us with Everett.

We part ways across the lake. Kathie, a prison puppy counselor and puppy-raiser friend who lives in Ludington, will meet me at the dock on her mountain bike. She's invited me to stay the night.

"We need to spend the morning getting our laundry done," Steve said when I asked if they wanted to ride the ninety-some miles to Clare together. They're on their own. Clare for me tomorrow. And home the next day.

We board the *S.S. Badger*, a seven-story steamship ferry with the capacity to haul 600 passengers and 180 vehicles, including RVs and commercial trucks. After wedging our bikes against the side of the hold below decks, we climb steep metal stairs and snag a table in the aft end lounge of the lower deck. The ship is not air-conditioned. We sit like sweaty sardines in a coal-fired can, eat, and settle in for an easy four-hour trip.

Except. One hour later we still haven't left the dock.

"Attention passengers," loudspeakers announce. "We are very sorry for the delay." A collective groan drowns out details about some kind of technical issue with the lifeboats.

Steve motions to me. "Let's step outside for a minute."

The shady starboard side is not so oppressive. Steve slumps his forearms against the gray iron railing, high over the water. The vast expanse of Lake Michigan is all that separates me from my home state. I peer to the blue horizon and wait.

"I'm having some trouble." Steve's concerns hit the metal deck like a gunnysack full of rocks. "I've been bleeding from my rectum for a few days. Remember when I had trouble with the heat? I called my doctor. I need to pick up a prescription in Ludington."

"You know, Ludington isn't too far from home for me. We have a van. I can call Andy to come and help you and the boys." They only have a couple weeks to get to Washington D.C.

"We'll be fine. I have a worker that can pick us up if we need it." He too studies the horizon. Words spill—his business, his off-season teaching job, his wife, his son, his life.

I cannot pick up his bag of rocks, I can only listen.

At long last the *S.S. Badger* sails. Back at the table with Everett, Matt takes advantage of Steve roaming the ship and fills his ears with music. *Enjoy the peace, my young friend.* I turn to the smiling boy.

"What do you think about your trip?"

I didn't think Everett's smile could get any wider, but it does. He tells stories of conquering mountain passes in the rain: "We didn't have any cell coverage one day and my mom freaked out and called the state police. We got pulled over at the top of the mountain. She didn't believe I was fine."

"We didn't have cell phones when I started touring. I called collect from pay phones and hoped my folks would accept the charges. On my first trip right out of high school, I didn't call home for a week. My mom was frantic, worried I would commit suicide or something."

We laugh. He gets serious. "Overall, this ride has changed me. My previous me was distrustful. Today's me realizes that good people are everywhere."

I hear my forty-years-younger me in his words, a cynical twenty-year-old, suspicious of adults over thirty, disgusted with my "conforming non-conformist" peers. I felt outside of everyone. B'76 opened my eyes. Beyond learning firsthand about geography, traveling by bicycle stripped away the strains of society, allowing strangers to meet one-on-

one on even ground. I came to view that people are good at their core, more alike than different.

"I love your term, 'previous me'" I say. "I felt the same way in 1976. People that summer told me, 'You'd better do this when you're young, because you won't be able to later.' Life and jobs and relationships happen. I'd like to go back and say, look at me now. It might be forty years later, but I'm out here doing it again."

"I'd like to do this again. What are you going to do next?"

My smile enlarges to match his. "I need to write a book."

"The story of your trip?"

"Yes, and more than that. It'll be a memoir of sorts. A nostalgic remembrance of my '76 ride, about my family, maybe other rides and experiences. I've done a lot of living in the last forty years." I am so comfortable with this kid, this kid who is like a mirror of myself, that I tell him about my Forrest Gump ride and how this ride came to be. "It isn't the ride of my dreams, but it might be the ride I need."

"How so?"

I open. I talk about Dad's death, how I've been grieving, how this ride is as much an inner journey as an adventure. Somewhere along the road, I let go of "riding until I am done" and focused on riding home. When Forrest Gump stopped running, he said, "Mama always said, 'You got to put the past behind you before you can move on.' And I think that was what my running was all about."

"I want to read it."

"You do?"

"Yes, but not the edited, finished version. I want to read your stream-of-consciousness writing. I want to see if you've felt what I've felt on my ride."

Oh Everett, I have very much felt what you are feeling. Back in 1976. Whatever will you think of my 2016 inner journey? I make a private vow to write for this young man, my previous-me doppelgänger.

Steve interrupts. "Come on up top. It's a lot cooler."

We are nearing Michigan and the last pedals of my long journey. Everett and I follow Steve. I'm happy for Matt, left behind with his music. We snake our way past passengers in the TV lounge and squeeze down the narrow hall that travels the length of the ship. Private staterooms line the outside port and starboard walls; restrooms, quiet room, museum, gift shop, movie lounge, kids' playroom, and video arcade fill stifling interior spaces.

"Just wait 'til we get to the stairs," Steve says. Near the front of the ship we turn left out of hell into a blast of heavenly frigid air. "Amazing, isn't it?"

"Can't I stay right here?" He motions me on. I grab the handrail to pull myself up the steep, steel stairs to the top deck and the open bow. And there is Michigan, land of pleasant peninsulas. My heart leaps. *No place like home? I won't even need a map from here.*

The captain twirls the behemoth ship around to float to the dock like a feather, the engines roar, and we kiss the shore. There's Kathie, snapping photos of us like we are celebrities. I observe myself as if I'm playing out the last scenes of a familiar movie, no longer master of my journey, feeling already home, wanting to delay the end.

I hug my three compatriots goodbye. "Save travels. Love you guys." I pedal away with Kathie. As I lay my head in her guest room, I ponder what I am about to accomplish, my there-and-back again adventure, a passage from past to future, a riding out of grief, an affirmation of self and Bunny-nature.

In three mornings I will wake up next to Andy.

STATS
31.51 miles, max speed 26.5, ride time 2:51, avg speed 11.1, TOTAL 3643.26

My previous-me doppelgänger and me, Michigan filling the horizon.

64

Center of my Universe

Thursday, August 11, 2016
Ludington to Clare, Michigan

briar patch

before rising
i feel
ribs, spine, scapulas against the earth
fingers wrap around my wrist
and overlap, this tight
foreign body is still
the body i have ever known

i am bunny

creator of my own life, I am
connected
stardust and energy
bones and thought

i am bunny and bunny is me
we are
trickster, lover
still, zig-zag
foreign, familiar

bunny is there and i am here
at once, one observation changes everything
when we notice
the other becomes
unnoticeable

be your nature, bunny tells me
stop noticing, i tell bunny

Such lovely humidity on this second-to-last leg. My legs, tight and sluggish, pedal through honey. It seems forever before I can announce "ten miles," another forever to twenty. Finally, after thirty miles, my self-audited ten-mile segments start to click. Fifteen minutes after noon I yell "fifty!"

Before jumping onto the paved Pere Marquette Trail that will carry me all the way to Clare, I divert into Reed City in search of Gatorade. A party store tricks me into adding a slice of pepperoni pizza. Nine weeks on the road and I still can't shake the habit of checking Facebook while I eat. Look at that, a message from my puppy counselors, Ron and Karen. They live in Auburn, not far from Midland. "Are you taking the bike trail? You are welcome to spend the night with us."

Funny, since my stay with Karen in Eau Claire, excitement for my journey spread throughout the Leader Dog family. The Pere Marquette ends in Midland, thirty miles beyond Clare, where I'll turn north. There was a time I wouldn't hesitate to add an extra day for a visit. It isn't today.

"Would love to stop by but I'm heading for Clare today and straight home tomorrow."

I wipe grease from my chin. A few moments later, a text from Karen. "You have a room at the Doherty. Paid in full. Our treat."

What? Too generous. "You deserve it," they reply when I send heartfelt thanks.

Expert at delay, I call Andy. "You won't believe this. Ron and Karen bought me a room at the Doherty tonight."

"Another night in a bed. A helpful transition to reintroduce you to life in the patch," he says.

"We'll see."

The Pere Marquette removes worries over heavy traffic out of Ludington. Grades are easy. A white-tailed deer freezes in the middle of the path. It doesn't move until I'm within twenty feet, then bounds ahead for seventy-five yards before darting off to disappear in nearby woods.

More than twenty-five miles later I am as crispy as the dry grasses that crackle at trail's edge. My pace snails. *Is the grade rising? Did the wind pick up?* I fantasize about Gatorade buried in barrels of ice. Like a granted wish, I spot the backside of an IGA at the end of a well-trod path. Red Gatorade and Cheez-Its—my diet is going to hell.

If I sit at the picnic table at the IGA any longer my head might drop into a nap. I force myself to move, I pedal like a slug. Going to be a long fifteen miles to Clare. "Air conditioning tonight," I bribe myself.

The ring of my phone interrupts crabbing. I'm so slow I don't even need to brake to stop and answer it.

Andy. "What do you think about me driving down to take you out for dinner?" It takes a moment to decipher what he said. "You have to eat, right? I won't come unless you say yes."

"Wait, what? Yes, of course! Will you stay the night?"

"Well, I have Gus to deal with. I'll have to go back for him."

Ah, well.

My legs rise from the dead, I average 13.5 miles per hour into Clare. I let Andy know I've arrived. Before hanging up I ask, "Can you bring me a shaver?" The razor I brought broke long ago. No need to bother shaving my Bigfoot legs until now, I have an unexpected date.

The front desk calls when he arrives an hour and a half later. I walk out to meet him. His familiar "hands swinging at his sides slightly slumped" walk flushes my face when I see him. He sees me. "Where's my wife? You look like a fourteen-year-old kid."

Sweet embrace.

He hangs out in the room while I shower. And shave. No more details, those over sixty can well imagine. Still, it's delicious.

We stroll, our fingers intertwined, to the Blind Tiger Pub and Eatery across the street. Neither of us hear the din echoing to the high tin ceiling, we are the center of the universe. He tells me about his recent trip to his brother's in Maryland, his niece's wedding, and a sad visit on his way home with elder friends who moved to assisted living.

"Linc recognized me when I walked in. His eyes lit up, but he couldn't communicate. Sally wasn't much better. I think she's depressed."

Without warning, my landmark 109-mile day east from Bismarck breaks out in tears and words. He holds my hand and listens.

"You need to write about this."

What did we eat? How long did we sit together? This last night is not at all what I expected: my tent a burrow, with the ground beneath me.

I am more than almost home.

STATS
91.06 miles, max speed 29, ride time 6:42, avg speed 13.5! TOTAL 3734.32

Andy and me, outside the Doherty Hotel.

65

Embracing Sunset

Friday, August 12, 2016
Clare to Lupton, Michigan *and HOME*

Passage

These hands with earthworms veining blue beneath gnarled bark,
Whose knuckles callous against the dying night,
Whose fingers rebel against the sun like fronds,
These hands with canyons dawning stick figures in the dark,
These hands, which slap like duck feet in tidal storms,
Whose skin sands the ages and laughs at larks,
These hands, which mold like powered metal presses,
Whose caresses fly with dreaming eagles,
Whose bones bend like elms in the winds,
Whose scars scream silent on pointed triggers,
Whose nails bite and groove the earth,
Whose palms are grinning Cheshire cats,
These hands, which rabble-rouse dust amidst the bedding,
These hands, which mold sorrow into snow to greet the spring,
Whose pain is harping rain on steaming asphalt,
These hands, which wrinkle in time and sink anchors into souls,
Whose sparking band is the universe singing eternity,
These hands embrace hands and cherish them
Like downy wings around a treasure.

Last day on the road.

Is Andy worried? Does he hope I've gotten this out of my system? I think back to watching a storm blow over the mountains surrounding Ovando with Jimmy, the Great Divide rider. "You are not done," he said with conviction. Can a stranger know me better than myself? Or did he see himself in me?

Keen to be home, I tuck unrequited wanderlust into grandmother's dusty attic trunk. I am grateful, previous-me-girl, that you embraced a drive to ride. Somehow you knew I'd need this, my "true" Forrest Gump ride. What is a "true" Forrest Gump ride? Gump ran across the country out of grief but running also freed him from bullies and restrictions. Throughout his life he ran for love and purpose, and his running created inspiration in others.

I see now my desire goes beyond a simple "ride until I am done." Riding is my own metaphor for living, for keeping a sense of curiosity, to witness others, of being in the now. (*Yes, Bunny, I remember.*) In his later years, our trepidation years, Dad would say, "Don't worry about me. I've had a good life." I've had a good life too. For now, I am content. Content with my dear in our bit of heaven, our patch.

There could be worse things.

Mother earth hurtles through the universe with no regard to me. The words of Stephen Crane worm past Laurie Anderson.

> A man said to the universe:
> "Sir, I exist!"
> "However," replied the universe,
> "The fact has not created in me
> A sense of obligation."

My self will never occupy the exact same point again. It's enough that this is true. The Mighty Mac spanned the leap to adventure, and I crossed the lake horizon to embrace the sunset of my life. And what will I do from here?

Live it.

Keep moving. Breathe. Be there.

Be here.

I study my right palm. If I trace a rising path from the knuckle of my pinky to the end of my heart line, that is how far I rode from Ludington to Clare yesterday. My fate line aims today's ride to where my middle finger meets my palm, followed by a tiered route to the middle of my pointer. Pointing home, the sunrise side. That's how we roll in Michigan, the mitten state visible from space, using our palm to point where we are. Or where we are from. (Mom might insist I add my left palm to represent the UP.)

"Ramon" rewinds with volume raised. Bunny sings along, intentions altered. *"Are you no one? So you've fallen. Light travels with you."*

Conscious always of my destiny (*Did you sense me with you Dad? Does Mom know of my promise to gather her in my arms when the time comes for her to leave?*), I need not know where we come from and what we are. I am of Michigan. I am my hands.

I travel still.

But before I go, breakfast. My room includes the Doherty's buffet. In bike clothes now two sizes too big, I am conspicuous in the hotel's fancy dining room with its white cloth-covered tables, fine china, and silverware. Until I smell bacon. *Bacon!* I care not. I pile my plate as if I have more than eighty miles to ride today. Oh wait, I do. Scrambled eggs, poppy seed muffin, coffee, water with ice, banana, and an apple snagged for later.

Cruising north through town on Business Loop US 127, I reminisce when Mom and Dad and I and my three younger siblings lived here in the early 1970s. Much is the same. Much has changed. The Mill End store, with its wood beam floor, is long closed; Stephenson-Doherty Funeral Home is now Stephenson-Wyman Funeral Home; the Whitehouse Restaurant still serves greasy burgers in its neon-lit box of a building downtown, but McDonald's, Wendy's, Arby's, KFC, Big Boy, Burger King, and Taco Bell replace the Ridderman farm at the north end. Fast food for the fast exit off the freeway.

The highway skirts Clare now. Back in the day, all traffic crossed US 10 at the town's self-proclaimed "Gateway to the North." We joked that if we stood at the corner on a summer weekend, we were sure to spot someone we knew driving through. I chuckle. What were the chances of meeting a Clare High School classmate in an RV park in Augusta, Montana?

A north wind. Only fitting as I battle hills on Old 27. No worries. I have legs and lungs now that I didn't have in June and a slow pace favors reflection. What have I learned on this journey back and forward? What lessons from Bunny?

Mostly, what I used to know. Hills level with commitment. Miles pass if you keep on. There is no shame in opting for the granny gear, walking, or accepting help, even when it comes unsolicited. Goals are met if you set to them with confidence. Trust experience. Trust your gut. You know more than you think, and your body can do more than you think it can. I *can* do whatever needs doing, even when resisting headwinds (real or metaphorical).

I also learned what I never realized I needed to know. Cauliflower doesn't keep well in the heat. There is redemption. And love pulls me home.

What will my life look like when I stop pedaling?

With Andy. Not returning to my job at the paper. I don't want to go back to the prisons, but I'm not done puppy raising. Strive to stay lean and fit. And I hope Mom doesn't need my aid in crossing over for a long while. She often says of her life, "There were lots of things I wanted to do and couldn't at the time. Now that I can do them, I don't feel like it." I suppose raising seven kids in a twenty-year age span tired her out, but her regrets likely fueled my pursuit of dreams.

This is my last chapter. I am eager to get writing.

East on Highway 61 (no hope of Bob Dylan taking over "Ramon" in my head), with just enough north-northwest wind to keep me out of my granny gear, just enough passing traffic to give me a pull, just enough shoulder for safety. Ahead the straight road narrows to a pinpoint but continues to open passage as long as I keep moving.

My cyclo-computer turns twenty miles as I cross the Gladwin County line. Forty years ago there wasn't such a thing as a cycle-computer, only a mechanical odometer attached to the front hub. It drove riders crazy with its clicks against the spokes. These days, you can start "Glympse," a GPS app on a cell phone. The app emails a website link to whomever you choose so they can follow your progress on a map.

"When you get within an hour of West Branch, can you start that tracking app?" Andy asked last night. "Linda wants to meet you on M-30 and Wickes Road." Linda plays bridge with us at the senior center on Tuesday afternoons and follows me on Facebook.

In the headwind fight heading north again, I daydream about Andy plotting a surprise party. *Isn't the Wagon Wheel Bar near Wickes Road?*

I imagine our Ogemaw County friends cheering me in, my boss from the paper snapping photos for next month's edition of the AuSable Voice.

A sharp gust beats me to my senses.

"Haven't learned enough, eh Bunny?" I smirk into the wind. Pedal on. Hills are less pointed, lovely forests fill the sky and open to fields of corn. Home. My mind empties.

M-30 takes a quarter-mile curve crossing into Ogemaw County. Two miles later a minivan stops at Wickes Road, zips across, and squeals a u-turn. Three people jump out to open the rear hatch. *What are they doing? Is that Linda?*

I coast.

Sure enough, Linda and another couple wrestle a refrigerator-sized cardboard sign around the van. "Welcome 'almost' home!"

Quick peak in my mirror, that'd be my luck, make it all this way and get creamed less than thirty miles from home. Clear. Before I can stand up off my seat, Linda pushes a cold water bottle into my hand and hugs my shoulders. Her friends are visiting from out of town. "We're following your Facebook posts, too."

We chat long enough for me to empty the bottle. No disappointment passing the Wagon Wheel Bar, my friend's heartfelt salute was plenty.

Momentum lost. It must be uphill to West Branch. There's the hospital. *Do I need to stop in?* Maybe I'm bonking. Instead I stop at the Forward Gas Station and Convenience Store on Houghton Ave, the main drag through town. Perfect. I've never been to this station, so entering feels on-the-road instead of back-in-town. One last Gatorade and cheese wrap sandwich is certain to make the last twenty miles easier.

Revived, I cruise past the familiar Victorian buildings of downtown West Branch, population a bit over 2,000, county seat of Ogemaw. I wave, to no one in particular, I doubt anyone takes notice of my passing. Left on N 4th Street with the Downtown Café on the corner, right on the stately residential street that carries me past Surline Middle and Elementary schools and onto State Street out of town.

Funny, I've never ridden State Street this direction. The northeast diagonal road heaves uphill in stages to M-33, reeking the sweet aroma of cows. Hills level on approach. My face hurts from smiling the way it did in 1986 when Dad and I rode the Michigan National 24 Hour.

At M-33, northbound weekend traffic (trunk-slammers, flat-landers, cottagers) forces my foot down. I almost forgot, it's Friday. Here State Road rolls east between acres of corn and farm crops, horse pastures, and cow lots. What's the name of that hamlet, little more than a handful of houses straddling the Rifle River?

Ah. There it is, the sign announces Selkirk with a forty-five miles per hour warning. Down, down, down to the river I coast to climb, climb, climb out.

"I'm not dead yet!" Turkey vultures soar above the Churchill Township Cemetery at the tippy-top of the hill.

"Good afternoon." Horses in a dusty paddock, hay hanging from their mouths, turn to watch me inch up another hill. "Haven't seen you in a while."

A turn off State Road onto Henderson Lake Road, which weaves through a forest. Two turns later and I head due north on Wiltse Road, five miles from home.

I have to pee.

Really? Can't I make it? The last mile, it's uphill, remember? Ugh. At the curve east, on the boundary of the Rifle River Recreation Area, I sneak into the woods. Relief.

Bring it on hill.

A half-mile later I turn onto Brady Road, the last mile I've dreamed about since leaving Missoula four weeks ago. One sharp quarter-mile climb before a gentle dip and the last grunt. At the top of the rise my phone rings from my pannier; rain threatened this morning. By the time I dig it out, I'm too late to answer Mom on FaceTime. I ring her back.

"Patti?"

"Hi Mom! I'm almost home! Look." I turn the phone around so she can see.

"I'm so happy you are safe. Now get going home to your husband."

We kiss the air between us.

The final three-quarters of a mile. Almost home, but I still have to get up this bloody hill. Truth is, this hill is no easier after weeks of building strength. No shame, I shift into the granny.

There—the patch. Mandy grips Gus's leash at the end of the driveway. He isn't sure if he should jump all over her or pull to greet me. Andy stands behind the mailbox with his point-and-shoot camera. I wave.

This man. He loves me enough to have watched me ride away. I love him enough to return.

POSTCARD FROM THE ROAD HOME 8/12/16

My tires sing as I lean a left turn onto Brady Road without slowing. Fitting that a north wind kisses my face welcome.

My momentum is not enough to carry me over the first rise on this last uphill mile. The field of corn east is sky high, growth as a passage of time. It's been sixty-five days since I pedaled my first mile west.

After a seventy-eight mile day at the end of a 3800+ mile adventure, the home hill is still a challenge. And I am glad for it. I want the returning to be worth working for, it shouldn't be easy, it must be desired.

A car screams by southbound. Without thinking I glance back with my helmet mirror. Another car approaches from behind.

"What is up with this traffic on MY road?!" I say aloud, forgetting it is Friday afternoon for the cottagers we call flat-landers.

"Hey! You are almost home!" the driver of the car yells, slowing beside me. It is Andy's youngest daughter Mandy and her friend, coming for a visit. They speed ahead and pull into our driveway at the peak of the hill. Mandy jumps out and runs toward the house; I am too far to hear her, but I imagine she is calling to her dad. "She's here!"

And I am home.

STATS

78.36 miles, max speed 33, ride time 5:39, avg speed 13.8, TOTAL 3812.68, return total 1790.19

Home hill selfie.

Afterward

Josh and Veronica made it too. Excited to see me arrive home after nine weeks on the road, they were also excited to see each other. I soon sat bunny-still, eyes hungry to look upon Andy (beaming in his chair) in our patch. I missed the glow of knotty pine walls with forest views at every window.

Wait. The table wasn't set. And the grill, I noticed it was still in the garage. Not like Andy to not have everything ready.

"Um, it's great I made it home to see you all, but…should I be starting dinner?" I was ravenous.

"Oh, we're not staying," said Mandy. "We're going to mom's tonight. We thought we'd give you and dad a night to yourselves."

"We'll barbeque tomorrow," Andy said.

The kids got the hint. And Andy took me to dinner at the Sunrise Café.

"When are you going to finish your book?" Mom asked a year after my trip. Thick into a second draft, I had no clue. Truth? I was nervous for her to read it. She didn't know about my abortion and even though she made no secret her aunt died from a self-induced abortion, I was sure she'd judge me. Then again, maybe she wouldn't. After Mom's admittance to hospice, my sister Anne and I shared shock when she confided, "One thing I regret, I was always so critical." (Is it just me, or does everyone flounder in personal conversations with their parents?)

I'll never find out. Before she died in January 2018 (and me, almost ready for a third draft) she said, "You'll finish it." I read my first chapter to her as she lay unconscious.

So strange to have this story bookended by my parents' deaths. If Andy worried I'd head back to the road to grieve Mom, he didn't let on. Perhaps the writing sufficed; editing is often more difficult than pedaling to Montana and back.

Mom's death was peaceful, unlike Dad's struggle. I think she was dying for months. Maybe she knew; I suspect she welcomed it. For her last Thanksgiving, my preacher brother and his wife (and dog) drove Mom to the patch. Anne and her three girls (and their dog) joined us. Mom sat on our loveseat with Jim's dog and petted ours (Gus, who avoids touch), smiling the whole time. Mom, who professed she didn't like dogs, was so gracious Andy and I wondered, "Who is that woman and what did she do with your mother?"

We are orphans now, my siblings and me. I miss Dad and Mom terribly. (Their ashes didn't end up in Lake Superior as Dad talked about. My brother Jim slid their urns together in their shared niche in Resurrection Cemetery in Clinton Township, MI.) Andy puts up with my smaller adventures, like sleeping in the snow at Tahquamenon Falls in the UP to attend a ski-joring clinic with my dog Aero. (No, Andy didn't come with us.) Or mentoring my young friend, Kayla, on a two-night backpacking trip on part of the North Country Trail. When I tell him I'll never leave him again like I did in 2016, he says, "You can leave, just promise to come back."

Andy knows me.

I will never be done. Never be done with living my life as art, dancing (my way) across the bridge between before and after, a personal performance of cosmic energy. I will never be done until life is done with me.

Appendix

Bike and Gear List

The weight of my bike and load? I didn't want to know. The thirty-five pounds I lost by departure date covered whatever excess I strapped to the bike. Upon my return, however, I weighed everything. Total combined weight of bicycle and gear was 105 pounds. Given I lost an additional ten pounds of body weight, my "load" once home was lighter than when I started out.

BIKE

I custom built my Tour Easy prior to 2004, when we still had the bike store. The Easy Racers size small, chromoly steel frame came with an Easy Racers stem, handlebar, and seat assembly. I added everything else.

- Shimano Ultegra 52/42/30 crankset with 165mm crank arms
- Shimano clipless pedals
- Shimano nine-speed barend shifters with an 11-32 rear cluster
- Shimano Ultegra front derailleur and XT rear derailleur
- Sram direct pull "V" cantilever brakes
- Sram nine-speed chain
- TerraCycle Elite Return Idler Kit for Easy Racers
- Custom-built wheels with forty-spoke Phil Wood sealed-bearing hubs, Mavic rims, Velox rim tape, and Schwalbe Marathon tires (20 x 1.50 front, 700c x 35c rear)
- CatEye wired cyclocomputer
- Red rear light attached to seat back
- Blackburn rear rack
- Zzipper fairing

A word about riding with a fairing. One would think it might help with head winds, but unless it was straight on, gusts made holding steady difficult. The fairing seemed to help maintain consistent speeds over 18 miles per hour, which wasn't very often. It did sweep bugs and a light drizzle over my head and proved invaluable in a hailstorm.

GEAR

New items I purchased included tires, a pair of rear Ortlieb waterproof panniers, an Aquapack waterproof camera bag for my Nikon, and maps from the ACA. The rest were things used on my first tours in 1974-75, the Appalachian Trail backpacking trip I took with Andy in 1994, or items I had on hand.

- North Face down sleeping bag rated to 20 degrees
- Tarp for under the tent
- Aluminum pot set and camp silverware
- Two ditty bags from B'76 for personal care items and wallet
- Shammy camp towel
- Thermarest short pad
- MSR WhisperLite stove and two Coleman fuel bottles (bottles carried in water bottle cages attached to the seat frame)
- Small, Teflon frying pan
- Stainless steel coffee cup
- Small bag for kitchen items (small spatula, GI can opener, spices, dish soap, sponge, towel, instant coffee, stove repair kit, lighter, waterproof matches, extra Ziplocs)
- Sierra Designs two-person, three-season tent
- Fanny pack attached to handlebars with a map case, knife, small notepads, pens, ditty bag wallet (driver's license, credit card, insurance card, cash), headlamp, etc.
- Small hydration pack (minus hydration) on seat back for tools and spare parts: two tubes for each tire, spare chain links and chain tool, tire levers, patch kit, tire gauge, miscellaneous nuts and bolts, spare shift and brake cables, T9 lube and shop rag, spoke wrench, multi-tool with Allen wrenches and screwdrivers.
- Clear map case strapped to handlebars
- Small bag with lip balm and sunblock
- Bungee cords
- Tire pump
- Four water bottles attached to stem and handlebar (I purchased a LifeStraw filter bottle in Wisconsin and recommend carrying

some kind of water filter.)

- ACA maps: North Lakes Bicycle Route Map 1, Northern Tier Bicycle Route Maps 4, 5, and 6, and Lewis & Clark Bicycle Trail Maps 4, 5, 6, and 8
- State maps: Michigan, Wisconsin, Minnesota, North Dakota, and Montana

CLOTHING AND PERSONAL ITEMS

I am forever grateful they invented Ziploc bags before my first bicycle tour. Invaluable for keeping clothes and food organized and dry, I packed each day's outfit in separate bags.

- Three sets of outfits including shorts, wicking t-shirts, socks, and underwear
- One wicking short and tank top (for swimming or when washing other clothes)
- One upper body base layer, one lightweight fleece top, one pair of lightweight tights
- Helmet, with mirror
- Cycling shoes
- Running shoes for off the bike
- Raincoat
- Bandana
- Prescription sunglasses and case
- Toothbrush and paste, floss sticks, shampoo, soap, razor, assorted pain relievers, allergy drugs, folding hairbrush, toilet paper
- First aid kit with sewing kit
- Bug spray
- iPhone 5 with charger
- FitBit with charger
- Nikon D700 with 40mm lens, spare battery, charger
- Wallace Stegner's novel, *Angle of Repose*
- Started with one medium-sized notebook, bought a second one on the tour, and two 3 x 5 spiral notepads
- Two pens

UP Pasties: A Taste of my Childhood

When our growing family was young, Mom made one giant pastie in a casserole dish, topped with a single layer of homemade pie crust. Rutabagas? Never. Dad wasn't a fan of the traditional ingredient of UP pasties. In later years, Mom returned to making single pasties; at some point she substituted her own crust with Pillsbury Refrigerated Pie Crusts. If they were good enough for my Yooper mom, they are good enough for me.

Mom's Pastie Recipe

Individual pasties (makes four or six servings)

Ingredients

Crust:
1 package of Pillsbury Refrigerated Pie Crusts, brought to room temperature (or make your own pie crust recipe of choice)
Filling:
1 pound of ground hamburger, raw*
1 small onion (I prefer yellow onions), diced*
2-3 potatoes (I like Russets), cut into small cubes, with or without the skins*
3 or so carrots chopped into small cubes*
Salt and pepper to taste
Assembly:
1- 1/2 tablespoons butter
Flour as needed
Small amount of water

Directions:

1. Preheat oven to 350 degrees.
2. Cut each pie crust in half (for four pasties), or thirds (for six, smaller pasties) and roll out on a floured surface to approximately a dinner dish-sized circle, slightly smaller if you are going for six pasties.
3. Mix hamburger, onion, potatoes, carrots and salt and pepper (these ingredients are uncooked at this point).

4. Place a good handful of the mixture on one side of each crust circle (1/4 or 1/6 the amount of mix) and top the mixture with 1/4 tablespoon of butter.
5. Dab water around the crust edge with your fingers. Fold crust over and pinch the edges together.
6. Cut two to three small slits into the top of the pastie for steam to escape.
7. Place pasties on non-greased sheets.
8. Bake at 350 degrees for one hour. They are done when the crust is golden brown and a fork slides easily into the potatoes and carrots.
9. Serve with ketchup (our family's Yooper way), or with beef gravy.

Pasties may be frozen (wrapped in foil and in a freezer bag) once cooled. Can be reheated in the microwave (one or two minutes on high) or the oven (set to 350 degrees, check every ten minutes). I use the microwave with good results.

*The amount of each ingredient in the mix is the same for four or six pasties. The percentage of meat to vegetables should be about equal. For variation, you could use cubed round steak instead of hamburger.

Acknowledgements

Does anyone read these acknowledgements? I do. Even before writing *Facing Sunset*, I realized the act of "finishing" (as in the context of carpentry) required the support and eyes of others. I have a long list.

First off, I thank my family, without whom I would not be the person I am. Mom and Dad, for life, for love, for accepting there was nothing you could do to stop me. Rick, for dragging me around on kid adventures, for step-sitting conversations, and for posting that Forrest Gump clip on my Facebook page as I neared home. Sue, for teaching me to tune my guitar, playing your accordion, and singing with me, for moving to the next block when your kids were small so I could be a fun aunt, for listening. Cathy, for illuminating the world with your art (and letting me use your sketch of Dad in Chapter 60), for caring for Dad and Mom, for respecting my words "as art." Jim, for riding with me in 1975, for the sixty-dollars you gave me on my birthday that I put toward panniers, for bringing a sense of peace into the world. Anne, for helping get flowers for mom and being there when she died, for being an unwitting nudge for me to move out on my own (that's another story, Anne, you might not have heard…), and to you and your girls for sending me the Tolkien-inspired wandering bracelet and for loving Rosie. Joe, for living long enough to enjoy retirement, for holding dad's hand, for telling me I am strong. Aunt Mary, for giving me a book about bicycles and touring so long ago.

My special helpers need special thanks. Tim Jenks and Susan Shantz, for letting me camp on your land during my test-ride, for the delicious pancake breakfast, and for the lovely journal that started this book. Debbie Bacal, for sharing our gypsy life before this life, for our Rover days, for investing in this ride (and my *Badger* ticket). Tammy Bartz, for feeding me and urging me, "It's just around the next curve." To Erika and Shane Hickey, for welcoming me like family. Kathie Bachman, for opening your home as I returned to Michigan. Karen and Ron Sharp, for the Doherty overnight.

And the ACA, for hosting the reunion that gave me an excuse to hit the road, for continuing the spread of touring by bicycle, and for permission to use a photo of their map. But I'll never forgive you for changing your name from Bikecentennial.

Without these early readers, including everyone who read and commented on my Facebook posts during my ride, I would not have had the courage to continue. Karen Voss, whose inciteful early read urged me on. Deb Donnelly, for telling me that writing my story was important work and insisting I add more about my experience with the prison puppy program. Katie Andraski, for her to-the-point questions, and ideas to structure my chapter memories. Catherine Sewell, for our lunchtime chats at the Sugar Bowl (or in the horse pasture). John Burke, for his attention to details. Debbie Bacal, for making me feel like I could write.

To those who read chapters pertinent to them, thanks for keeping me honest (and for more). In no particular order: Bruce and Susan Newells (for "paying it forward"), Sandra Moeller, Michael Prest, Erika and Shane Hickey (that you listened, Shane, means more to me than you'll ever know), Donna and Scott Reynolds (for friendship), Melissa and Miguel Torres (for the jump!), Rich Landers (for dinner), Dave Lindstrom (for sharing the reunion), Verna Ann Wyke (for "girls' night"), Sharon Stoick (for riding with me), Linda Scott (for your suggestion of the title "Gone With the Damn Wind" and welcoming me "almost home"), Dave Marzolf (for busting me), Joseph (JP) Flood and Jane Lawther (for helping rid me of "shoulds"), Hunter Lydon (for the salve and campsite), Bob Downes, Mary Bergkamp-Hattis (for tempting me with your husband), Robin Ring, Loren O'Brien (for getting me to the Honey Hub), Kathie Bachman, Karen and Ron Sharp, Linda Boston, Mike Dobies, Rick Brehler (for letting me use some of your words), Catherine Peet, Jim Brehler, Joe Brehler, Anne Bicego, and Marcia McConnell Hale.

To all my future readers: dare to be bold!

Without these dear friends, my life would not be so rich. Robin Ring, for agreeing to ride with me in 1974, just "because you asked me!" Mike Dobies, for all our training rides and for offering a wheel when I needed it. Linda Boston, for your art and our time at Kwagama Lake. Lon Haldeman and Susan Notorangelo, for pushing me to ride further. Karen Voss, for our special connection and standing witness with me.

And Lou Hotton, my evil twin, for always having my back and

insisting I do my best, for telling me, “Run as fast as you can to that tree up there and jog back to me,” for synching on our tandem. Everyone should have a friend like you, who dreams big and gets moving.

Sincere thanks to my online writing group, The Writing Herd. Without their enthusiastic support when I shared with them the hardest chapter in my story to write, I never would have gotten beyond writing just for me.

Ah, where would writers be without editors? Thank you to Elisabeth Kauffman, of Writing Refinery, for her professional overview of draft number five; to Angela Mac, whose communion over my words done sooner in my writing process might have prevented so many drafts; to MaxieJane Frazier, of Birch Bark Editing, for her optimism and expertise in drawing out my best.

Special thanks to my artist niece, Sofia Bicego, for illustrating my west and east bicycle wheels.

Anna Blake and Crissi McDonalds’s Lilith House Press deserves credit for raising women’s voices. In another bit of synchronicity, Anna Blake started The Writing Herd shortly after I returned home from my ride. Perfect timing. Thank you, Anna, for saying “Yes!” For suggesting I start another rewrite in the middle, for pointing me to professionals for help, for kicking me in the pants when I needed it, and for housing me in your tack room, your lovely old touring bike hanging over my head.

To all the people I met on my trip, thank you for reminding me, during times of unrest, that we are all in this together and at our core is love. Colleen Hueffed, for driving to Ovando and feeding us. To the couple in Wisconsin who drove me to Hayward, I am sorry I didn’t get your contact information—thanks for the ride. Rick, in Sentinel Butte, for sharing your table. Robin Heil and Loren O’Brien, for reminding me where I started. Everett, my doppelgänger, for your smile and openness to the road. Santosh Ramdam, for recognizing me. The mare in North Dakota, for speaking to me. The hawks, for lifting me. And, of course, Bunny, whose appearances came at opportune times, for acting as my spirit guide.

There is no room for the rest of you here, but you are in the book.

And finally, I am indebted to Andy, for his unwavering support, his enabling, for keeping the home fires burning. What a lucky girl I am!

Author Page

patti brehler is retired as a freelance writer, photographer, and editor for a small-town newspaper. *Facing Sunset* is her first book. She is one of thirty-three contributors to the anthology of women's voices, *What She Wrote*, (published by Lilith House Press in 2020), and has essays published in the *Massage Therapy Journal* and the *International Association of Assistance Dog Partners.*

patti lives in her beloved "patch" with her husband, Andy Andersen, and their two black labs, Gus and Aero ("career-changed" from Leader Dogs for the Blind). Her new adventure? During the 2020 Covid-shutdown, she bought a horse—a tri-colored Overo paint mare, aptly named Crazy Horse. Find patti on Facebook at https://www.facebook.com/patti.brehler. Her email is pattibrehler@gmail.com.

Made in United States
Orlando, FL
20 November 2022

24771494R00235